Cape Wind

Cape Wind

Money, Celebrity, Class, Politics, and
the Battle for Our Energy Future

Wendy Williams

and

Robert Whitcomb

PublicAffairs
New York

Art Credits
Line drawings by Nancy Spears
Photos by Gregory J. Auger
Kennedy image courtesy of Greenpeace USA

Printed in the United States of America.

Public Affairs books are available at special discounts for bulk purchases in the United States by corporations, institutions, and other organizations. For more information, please contact the Special Markets Department at the Perseus Books Group, 2300 Chestnut Street, Suite 200, Philadelphia, PA 19103, or call (800) 810-4145, ext. 5000, or e-mail special.markets@perseusbooks.com.

Library of Congress Cataloging-in-Publication Data

Williams, Wendy.
Cape wind : money, celebrity, class, politics, and the battle for our energy future on Nantucket Sound / Wendy Williams and Robert Whitcomb.
p. cm.
ISBN: 978-1-58648-397-5 (HC)
ISBN: 978-1-58648-575-7 (PBK)
1. Cape Wind Associates. 2. Wind power industry–Massachusetts–Nantucket Sound. 3. Energy policy–Massachusetts–Nantucket Sound. 4. Renewable energy resources–Massachusetts–Nantucket Sound. I. Williams, Wendy. II. Whitcomb, Robert. III. Title.
HD9502.5.W554C37 2007
333.9'20916346–dc22

2007001106

10 9 8 7 6 5 4 3 2 1

From Bob,
To Nancy

For Jane, Denley Ann and Brett and Jenny,
Abby Jane and Niki and Neily

I, your servant, will build a mill which does not require water,
but which is powered by wind alone.

Leonardo da Vinci

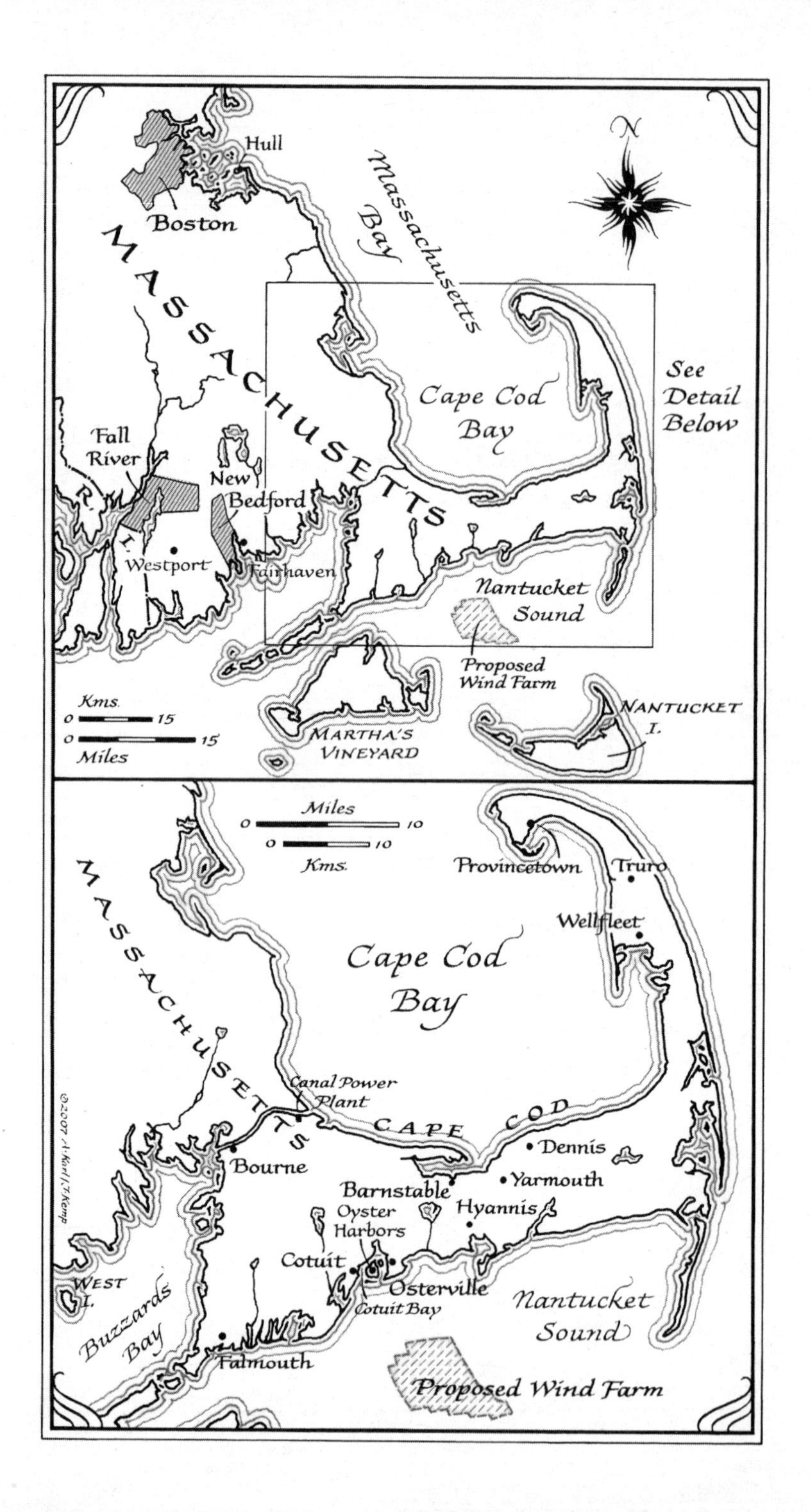
Hull
Massachusetts Bay
Boston
MASSACHUSETTS
Cape Cod Bay
See Detail Below
Fall River
R. I.
New Bedford
Westport
Fairhaven
Nantucket Sound
Proposed Wind Farm
Kms.
0 15
0 15
Miles
MARTHA'S VINEYARD
NANTUCKET I.
Miles
0 10
0 10
Kms.
Provincetown
Truro
Wellfleet
MASSACHUSETTS
Cape Cod Bay
Canal Power Plant
CAPE COD
Dennis
Bourne
Yarmouth
Barnstable
Hyannis
Oyster Harbors
Cotuit
Osterville
Cotuit Bay
WEST I.
Buzzards Bay
Nantucket Sound
Falmouth
Proposed Wind Farm
©2007 A. Karl / J. Kemp

CONTENTS

PROLOGUE

A "Public" Meeting

Democracy! Bah! When I hear that
I reach for my feather boa!

ALLEN GINSBERG

December 6, 2004
Martha's Vineyard Island, Massachusetts

David McCullough's face contorted with anger. "It's outrageous!" he yelled. He sounded like anyone but the mellow, well-measured man of letters who narrated tales of American history on national television. Indeed, the popular author sounded quite overwrought.

Nantucket Sound, he shouted, is "hallowed ground." He had uttered that phrase before, as the narrator of Ken Burns's famous Civil War documentary aired on public television. This time, however, he was sounding his own battle cry, crowing his promotion to general in the seaside civil war, a war that had become an internationally watched conflict over the future of energy and of America's air, coasts, and oceans.

On this late-fall evening, as the sky spit a chilly mixture of snow and sleet, McCullough's big voice filled the high school auditorium lobby on Martha's Vineyard, an island favored by movie stars, politicians, the international jet set, authors, and other glitterati.

He continued his mini oration, "This is a preservation issue. It's not an environmental issue."

McCullough, surrounded by a small circle of admirers, had just walked out of the first of a group of four public hearings called by the U.S. Army Corps of Engineers regarding a proposal to build a large electrical-generation project in the middle of a body of water known as Nantucket Sound. Cape Wind Associates, a limited liability company, wanted to place a vast field of wind turbines out in the salt water. The turbines would, said the developers, produce enough clean electricity to satisfy a considerable proportion of Cape Cod's needs. Indeed, on rare occasions, the project would supply all the necessary electricity, obviating the consumption of fossil fuels.

Over the three years the battle had raged, McCullough's opposition to the project had become common knowledge around Cape Cod and the wealthy islands of Nantucket and Martha's Vineyard, in part because of a highly emotional radio advertisement he had recorded that excoriated the project.

Rarely, however, did McCullough appear at public meetings about the wind farm. Indeed, like so many of the beautiful people engineering the public show of fury over Cape Wind, the television star and author preferred controlled, closed-door situations. Tonight, though, after refusing a request for a formal interview, he sputtered on.

"This is visual pollution," he complained. He was unable to stop talking. As he and his entourage departed the building, the sentences trailed after him, like leaves blowing in a high wind.

David McCullough voiced his opposition to Cape Wind from the earliest days of the seaside civil war.

Cape Wind, the brain child of a small group of innovative energy developers, first made headlines in the summer of 2001, only weeks before September 11. The first adamant public opposition surfaced less than a month after the World Trade Center disaster. Cape Wind president Jim Gordon, a fiercely driven and ardently independent man who had made millions during his thirty-year career as an energy entrepreneur, had initially proposed erecting 170 towering wind turbines five miles off the Cape's southern coast. During the course of the battle, as offshore wind technology improved, the number of turbines had dropped to 130, but the output of the individual turbines had grown considerably, from 2.8 to 3.6 megawatts each. The project's resultant "nameplate capacity" would have been 468 megawatts, quite large for a wind farm. The project's typical output, however, would have been rather less, since wind turbines only rarely operate at full capacity.

Because Nantucket Sound's winds often blow best when electricity is needed most, during the frigid wintertime when fossil-fuel costs are high or on hot summer days when Cape Cod's many 7,000-square-foot shoreline homes have their air conditioners pumping hard, an ocean-based wind farm seemed to Gordon an obvious solution to New England's power-generation dilemma. Lacking indigenous fossil fuels, the region suffered extremely high electricity prices. Moreover, much of the region's electrical generation was aged, inefficient, and consequently highly polluting.

Because New England's coastal regions are already so overdeveloped, little space remained for land-based wind projects. Building offshore seemed an obvious solution to the crisis. Nevertheless, Gordon, a wildcatter, had shocked the region by proposing the massive project. Although offshore wind had a successful ten-year history in Europe, the projects to date had been relatively small-scale. Nothing this ambitious had been built, although several proposed projects far exceeded the size suggested for Nantucket Sound.

Gordon was undeterred. Ebullient and confident, he believed his project would pay off financially while also cutting down on fossil-fuel emissions. From Gordon's point of view, the region would be trading a small area of the ocean, used mostly for recreational sailing and saltwater fishing, for cleaner air and a leadership role in clean-energy innovation.

Cape Cod's powerful elite saw things differently. To them, Gordon and his team were interlopers. The same winds that had enticed Gordon to gamble his millions had, as early as the nineteenth century, enticed large flocks of the wealthy to nest along the Cape's hitherto rather neglected southern shoreline. By the time Cape Wind made headlines, the area of Cape Cod, Nantucket, and Martha's Vineyard had become a devil's triangle of entrenched, often inherited, wealth. Those seashore homeowners had come to see Nantucket Sound as their very own playground.

Few Americans will ever literally *see* Nantucket Sound. Most of its beaches are intentionally closed to the public, who are instead directed further toward Cape Cod's eastern end, to the national seashore that overlooks the Atlantic Ocean. Still, many Americans have "seen" Nantucket Sound on television. These are the waters enjoyed by Jack and Jackie Kennedy, who sailed out from the summer capital of their brief Camelot—Hyannisport. The carefully orchestrated, nationally broadcast images showing the romantic young couple sailing their elegant little boat across the glittering waves brought about the end of "quaint" Cape Cod. Already drowning in money, the southern shoreline, now made famous by the Kennedy family, was swamped by a storm-surge of development.

Today, much of Cape Cod is a highly commercial, Disneyesque version of what was once a very lovely seaside area. Along the south shore and on Martha's Vineyard and Nantucket, colonies of the rich and the hyperrich flourish as never before—not just multimillion-dollar folk, but *really* rich people—such as Jack Welch, once leader of General Electric; Paul Fireman of Reebok; Douglas Yearley, longtime chairman of Phelps Dodge mining and board member of Marathon Oil Corporation; and William Koch, inheritor of money from Koch Industries, a massive, privately held energy company heavily into fossil fuels. Like Fidelity's Abigail Johnson, ranked America's twelfth richest person in 2005 by *Forbes* magazine, some represent financial money. Some, like Koch, come from purely industrial wealth. Many, like the Mellons and the DuPonts and the Kennedys, have been there for decades. During the summer, Nantucket Sound can be a busy crossroads—except on Horseshoe Shoal, where Gordon wanted to put his wind farm. There, with only a few feet of water at low tide, it's too shallow for most yachts.

While McCullough fulminated in the lobby, a Category 5 hurricane raged inside the tiny auditorium. The Corps' hearing was supposed to be a kind of New England town meeting, allowing the public to comment on a 3,800-page, $10 million Environmental Impact Statement, three years in the making. Speakers, whether just local homeowners or world-renowned scientists, could come to the microphone to discuss the document's content. Hearings like this are usually routine, tedious, and filled with coma-inducing facts and statistics, leading the junior reporters stuck covering them to wonder how early they can leave without jeopardizing their jobs.

Tonight, though, a treat awaited the bored press corps. The hearing had been hijacked. Science, statistics, and facts were no longer on the agenda. For that, reporters could thank U.S. Congressman William Delahunt, his longtime aide and political fixer, Mark Forest, and the Alliance to Protect Nantucket Sound, a shadowy group claiming environmental credentials but backed big time by fossil-fuel investors, members of the animal rights crowd, an ancient arbitress of high society, several Bush Pioneers, an athletic-shoe maker, high-tech moguls, investment bankers, real-estate developers, and a long list of donation-savvy politicians.

Heading up the Alliance and bearing the mighty title of Chief Executive Officer was mining and fossil-fuel baron Douglas Yearley, who owned a $6.8-million, 7,700-square-foot house on Nantucket Sound. From his property he would easily see the wind turbines on clear days—a prospect that "offended" Yearley, according to his public-relations man.

To the casual observer it might have appeared as if Vineyard residents were unusually interested in the wind-farm issue, but in fact the Alliance had shipped over many meeting-goers via boats. Arriving at the high school, these ringers huddled in lobby corners before the meeting, waiting to be coached in the art of meeting manipulation by Susan Nickerson, the $90,000-a-year Alliance Executive Director, and Ernie Corrigan, her well-paid hatchet man.

Other Alliance workers, a few of whom had been added to the payroll for just this occasion, barked out orders. Some handed out free bottles of water and cupcakes with red-and-green sprinkles, along with glossy Save Our Sound fliers. Some deluged the speakers' sign-up table, trying to get in early in order to monopolize the registration and speaker list, overwhelming the harried and stunned U.S. Army Corps of Engineers employees. Beleaguered Corps staff even had to fend off insistent efforts by some Alliance supporters to combine the times scheduled to be allocated among several speakers so that one wind-farm foe could make a major speech.

Outside the auditorium, in the chilly dark, other Alliance importees enthusiastically chanted, "The Sound is not for sale!" Paid staffers carried signs. Little round stickers on their coats and sweaters also cried out Not for Sale! When these people spoke at the microphone, many would provide their names but not their addresses, giving the impression that they were island locals instead of imports.

Strutting around like a ringmaster was the Alliance's PR czar, Ernie Corrigan, dressed jovially in pre-Christmas red shirt and green tie. Corrigan, a beefy man with a thick shock of graying hair and a belly just beginning to hang out over his belt, was, as a younger man, a reporter for the nearby New Bedford *Standard Times.* For a short while he covered the Massachusetts State House for a chain of small dailies owned by Dow Jones. Then, fancying himself a kingmaker, he began managing small-time elections around the state.

Corrigan looked in his element as he directed and hovered, chatted with journalists, and whispered into his cell phone. He glad-handed as if his life depended on it. He was the picture of energy, a man in command.

This was fun to watch, but it was really just a sideshow to the evening's pièce de résistance: the presence of the Honorable William Delahunt, the white-haired U.S. congressman whose district included the Cape, Martha's Vineyard, and Nantucket, as well as towns and cities closer to Boston, like working-class Quincy, whence Delahunt hailed. Delahunt hated the wind farm. Or, at least, he *said* he hated it. Delahunt was widely seen as Senator Edward M. Kennedy's man. What Ted

Kennedy hated, Bill Delahunt hated. And Ted Kennedy loathed Cape Wind, with an unwavering ardor that curiously belied the environmental ideals he so often proclaimed from the floor of the U.S. Senate. Nantucket Sound, Delahunt repeated, is a "precious resource."

Ironically, Delahunt was fighting a proposal that promised to help his own economically stressed hometown. One proposed assembly site for the massive wind turbines, whose parts would arrive from all over the globe, was a closed-down Quincy shipyard. The closing of the yard had devastated the city. Its reopening would give unemployed Quincy workers many highly paid jobs.

As his boss raved at the podium, Delahunt aide Mark Forest, known in the district as "the Little Congressman" for his longtime political reign on Cape Cod and the Islands, leaned against the auditorium entranceway. He watched as Delahunt took the podium and delivered nearly verbatim the lines he'd been delivering for three years. "Nantucket Sound is not our backyard, it is our front!" the white-haired congressman shouted.

"It is not just a view for those living and working on the water," he continued. "It is an economic engine. It is the heart and soul of our region. . . . Cape Wind does not own Horseshoe Shoal. Cape Wind has no right to use it and the Congress of the United States has . . . not dealt with the problem."

For most of the 300,000 or so locals who lived in the area but who rarely visited Nantucket Sound, the fact that this particular body of saltwater was the region's "heart and soul" was news indeed. To many, the exaggeration sounded simply silly. Nevertheless, project opponents, including the Alliance's evening immigrants from the Cape and the mainland, cheered and clapped repeatedly.

"Let's not lose sight of the fact that this is a public resource, that the waters and the seabed are owned by the American people," Delahunt continued.

Then began the requisite listing of useful environmental icons: whales, seals, birds, fish, and so on, whose existence the wind farm supposedly threatened.

"This is a public resource, owned by the American people, by the U.S.A.," he said, tripping over his own lines.

The congressman's face reddened. Like old Cotton Mather warming to his subject of sin, Delahunt's fiery ardor mounted. He curled his fists and began a bit of podium pounding. This time, Delahunt had decided, Jim Gordon, the man who refused to go away, would finally realize it was time to throw in the towel.

The congressman, a former prosecutor, promised "endless litigation"—"nothing will get done"—if the Army Corps of Engineers gave its approval to Cape Wind.

Forest, a small man who'd spent the whole of his political career on Cape Cod, was still leaning against the auditorium entranceway wall. His arms were folded, as usual. He was trying to keep his face passive, although a certain shine to his eyes betrayed his satisfaction at having maneuvered his boss up onto the stage.

Asked if Delahunt's words were meant as threats intended to deter Gordon from continuing to press his case, Forest edged away. "Speculation," the fixer answered curtly.

Delahunt's control of the podium was unusual. Every other speaker had to use a floor microphone and was limited to three minutes. To maintain discipline, a very large traffic light turned first a warning yellow and then a time's-up red. Delahunt, however, assumed he was exempted from the burden laid upon the rest of the hearing's participants.

Blindsided by the congressman's performance, project supporters—and there were plenty on Martha's Vineyard, despite the Alliance's efforts—were miffed. How had this politico gained control of what they thought was to be a "public"—as in, for the *public*—hearing, an opportunity for thoughtful and informed people to add their insights to the discussion? (In fact, Forest had forced Army Corps officials to bow to Delahunt's coup d'état. Had the Corps refused, the congressman could have taken out his revenge when appropriation votes came up on Capitol Hill.)

As the scale of the Alliance ambush became clear, project proponents, scattered throughout the crowd, looked more and more confused. Matt

At a U.S. Army Corps public meeting on Cape Wind, project supporters from Cape Cod let the world know that the Alliance to Protect Nantucket Sound isn't the only game in town.

Palmer and the Reverend William Eddy, cofounders of a nonprofit group supporting Cape Wind called Clean Power Now who had themselves come over for the night from Cape Cod, were babes in the woods when it came to this kind of behavior. The Alliance's slogan was Save Our Sound, with the emphasis on "our." Palmer and Eddy had coined their own motto—It's Not the View, It's the Vision—directed toward policy rather than possession. Big bucks and political power plays were alien to them. Political innocents, they believed there would, one day, be a reasonable discussion of the project's merits and that opponents would come to see the light.

Listening to Delahunt, that seemed doubtful.

Jim Gordon, the fifty-one-year-old target of this tempest, sat quietly by himself in the midst of the crowd, almost unnoticed. Over the course of the three years, Gordon had tried to perfect the art of public aplomb, but as Delahunt pounded the podium, Gordon began to scowl. The scowl deepened and darkened as the evening progressed. A careful observer could tell when Gordon was feeling hard-pressed. Always a man who sat ramrod straight, he sat even straighter with each of Delahunt's personal insults. From time to time he worked his jaw forcefully, trying to remain impassive.

Throughout the battle, he had heard similar rhetoric, some of it harshly personal. He'd been belittled, called names, sneered at as "a Boston developer," and repeatedly maligned as "greedy." His staff had been physically threatened. Articles intended to destroy his business reputation had appeared in the local daily newspaper. He'd been dragged through the court system to the tune of more than a million dollars. His supporters had been verbally attacked. In some cases, their careers had been threatened.

Although he struggled not to show it, this was deeply painful to Gordon. After nearly thirty years in the energy business, the man was no stranger to aggressive behavior, but a small part of him had genuinely believed that New Englanders—even Cape Codders—would welcome his proposal. Instead, he'd encountered a level of viciousness that went far beyond normal business tactics. Gordon had always been a rough-and-tumble guy, but to stay in the seaside civil war he'd had to grow an ever thicker skin. Yet despite plenty of opportunity to take on other projects, he'd refused to quit. Quite possibly, he just didn't know how.

Project opponents had missed a key aspect of Gordon's personality. For this man, money was a motive—but not the only motive. He was a businessman, but he was also a talented leader and team builder, a change agent and a groundbreaker. He saw himself as a man who freed good ideas from their Ivory Tower encasement and made them manifest. Three years of public excoriation had served only to deepen his determination. By the time of the Martha's Vineyard hearing, Jim Gordon, energy entrepreneur, had become a man on a mission.

Indeed, over the coming several years, Gordon would increasingly appear to be leading, for better or for worse, a crusade. At one point, several years in the future, while rallying his troops on the island of Nantucket during the height of a battle over legislation Kennedy tried to sneak through Congress, this multimillionaire builder of power plants would thrust his right hand up into the air and shout: "This is the way revolutions start!" He would come a long way from being just a power-plant guy.

But for now, for the time being, while listening to Delahunt rant, Gordon kept to himself. To those who didn't know him, he appeared frustratingly stolid, almost as though he had something to hide. Throughout the course of the battle, enraged wind-farm opponents had asked him over and over in public meetings, with increasing anger and impatience: Why are you still here?

And for the past three years, this self-made man and child of the 1960s had just shrugged his shoulders, chanted his mantra of "clean, renewable energy," and kept on truckin'.

The Alliance's parade of carefully scripted opponents lined up to play to the television crews, including one from Tokyo. Foes repeated the sound bites that Corrigan and other hired guns had made up so long ago. Gordon felt more disappointed than angry. Like so many others, including many of the undecided, he had yearned for an end to the political theater and for a fact-based debate.

That conversation wasn't to happen that evening. After Delahunt's performance was over, it became obvious to those who had expected a hearing packed with content that they'd been naïve. It seemed that the three long years of study and science, of trying to trust the democratic process, of hoping that the truth would come out, had been futile. These public hearings, paid for by taxpayers, had, like so many of the earlier ones, been reduced to little more than an expensive opportunity for political hit men to grab the podium and prove to their moneyed supporters that they were sticking up for them.

Yet even as Delahunt rallied his troops from the hijacked podium, the lawmaker neglected to address the issue at hand, the real-life reason for the public hearing—the Environmental Impact Statement assembled by the Corps. Par for the course, thought Gordon, who believed that the report itself had been positive for his project. No one had come up with any science-based reason that the project shouldn't be built. There had been no substantive safety concerns, no evidence that endangered species would be harmed or that the environment would be damaged, no real roadblocks of any kind. A variety of governmental and non-governmental organizations had asked for more information on several matters, as was typical, but the project on the whole seemed to have been given a scientific thumbs-up.

By evening's end, one thing was clear: This melodrama—Corrigan strutting in his Christmas best, Forest hovering around the edges, Delahunt vehemently pounding the podium, Ted Kennedy's influence ever felt but never seen—was a very well-orchestrated show whose paymasters had immense resources. It was all very much in keeping with the previous three years, during which the Alliance had been engaged big time in creating a parallel universe, a virtual reality, their own Hollywood storyline, about Gordon's project, about Nantucket Sound, and about the environmental and energy issues facing Cape Cod, New England, and all of America.

Reporters had fun for a while that evening, but upon reflection, some were saddened. The hearing was supposed to be an opportunity for public discourse and an expression of democracy at the local level. Instead, it had been hijacked and turned into a publicity stunt. While wrapping themselves in the mantle of democracy, the Nantucket Sound affluent were behaving as though they owned the government.

Jim Gordon's big idea was bold, breathtaking, even brash, just the sort of project that used to be associated with American entrepreneurship. But it was getting harder to introduce such ideas in a nation that had become increasingly dominated by an entrenched plutocracy that had little, if any, sense of national or global responsibility. America used to be a nation of bold ideas, but by the twenty-first century, some had begun to wonder whether genuine visionaries might do better in some other nation.

When a democratic process could be sold like this to the highest bidder, and when a U.S. congressman was present to do the honors, what did this mean for the future of America? A few of those present that evening found the symbolism of the event frightening, given the dangerous realities of the new millennium. Energy prices were steadily rising. Regular people were having trouble paying their bills. Climate change seemed to be under way. Oil and gas were in short supply and developing nations were eager to have all that electricity could provide, from lightbulbs to computers.

Somehow, somewhere, sometime soon, these challenges were going to have to be addressed—by someone willing to take the lead. Delahunt was funny, several reporters agreed, but many Americans didn't seem to be in the mood for such shenanigans.

"Nero's fiddle," muttered a journalist watching the show.

How had it come to this?

CHAPTER ONE

Jim Gordon's Big Idea

Three may keep a secret, if two of them are dead.

Ben Franklin

Summer 2001

The season began typically enough. Having just completed its annual plant and bake sale, the Osterville Garden Club prepared for its midseason garden tour. The Reverend David Angelica, of the town of Orleans's Church of the Holy Spirit, praised the miracle of millions of dollars raised to renovate the church sanctuary. Dennis residents donned yellow-and-white T-shirts emblazoned with Save the Crowe, a campaign to save some pastureland from what they called "the invasion of McMansions." All over Cape Cod, men raked leaves and clipped hedges, women planted flower bulbs, kids went to malls. The sound of traffic began to climb to its summertime crescendo.

Out on the ocean, Swedish linen exporter Bernt Larsson and his wife neared the Cape after sailing their forty-four-foot sailboat across the Atlantic. Government scientists had just completed a routine tagging of harbor seals off Chatham. The Coast Guard was searching desperately for a Yarmouth fisherman lost at sea, and staffers at Provincetown's Center for Coastal Studies worried about a right whale tangled up in fishing gear.

Down in Washington the Bush administration talked about opening up Georges Bank, off southeastern New England, to fossil-fuel drilling, outraging local environmentalists and fishermen, as well as Congressman Bill Delahunt's political aide Mark Forest. The *Cape Cod Times* wrote angry editorials opposing the plan and calling for clean-energy solutions.

Senator Edward M. Kennedy was complaining about the Bush administration's refusal to sign the Kyoto Protocol and was leading the charge against drilling in the Arctic National Wildlife Refuge, going so far as to vow to filibuster on the Senate floor. The senator did, however, find time for the Figawi Race, sailing from Hyannis to Nantucket, over the Memorial Day weekend. He was "honored" to participate, he said. In years past, the sailboat race had earned some pretty bad publicity when participants enjoyed a bit too much drunken revelry over on the island, but the race spokesman promised that participants these days were "more mature." Some Cape Codders may have believed him.

For another famous Cape Cod sailor, William Koch, of elite Osterville's yacht-crazed Oyster Harbors Club crowd, it was also business as usual. Koch, who had once won the America's Cup and who had lived long and hard on those laurels, was winding up yet another of his lengthy court suits. This particular legal battle had been waged against his brothers, who ran the family fossil-fuel business inherited from their father, Fred Koch, supporter of the John Birch Society and founder of Koch Industries. Bill Koch's theatrical legal fight ended by his giving in to his brothers, selling them his share of the company, and then starting

Massachusetts state representative Matthew Patrick, beside his hybrid vehicle, one of the first on Cape Cod. He nearly lost an election because he asked Cape Codders to stop to consider the project's merits before voting to oppose Jim Gordon's Cape Wind project.

his own energy company, Oxbow Group, which specialized in a variety of extractive- and fossil-fuel-mining endeavors. "Blood and money is a very explosive mixture," Koch had recently told CBS News.

Cape Codders always look forward to summer, but this year they were even more ready for the commencement of party season. The past winter had been nasty. While the nation had weathered the storm of the Bush-Gore election and the Enron saga, local electricity prices had skyrocketed.

"The high prices for oil are here to stay," warned Matthew Patrick, a forty-nine-year-old newly elected state representative who had once led a citizens' organization dedicated to reducing citizens' energy costs.

Cape Cod's many low-income and fixed-income residents clamored for solutions. On June 18, readers of the daily *Cape Cod Times* found an editorial touting the benefits of ocean-based energy technology. "Offshore wind and wave energy could supplement our power grid," *Times* editors wrote. Calling the ocean an "untapped powerhouse of energy," editors praised the German government for suggesting that offshore wind projects could replace nuclear-power plants.

Reading the editorial, Jim Gordon smiled to himself. He and his team, working in Boston sixty miles north of Cape Cod on their not-yet-made-public wind-farm proposal, were about to give the editors what they said they wanted. The paper frequently ranted about the pollution emerging from the fossil-fuel-fired Cape Cod Canal electrical plant, a mid-twentieth-century dinosaur that provided most of Cape Cod's power. Sometimes emitting burps of corrosive chemicals that destroyed paint jobs on cars and boats, the plant spread deadly airborne particulates over many miles. Not just Cape Cod suffered. Elsewhere, in the poorer cities and communities to the west and north, the air was bad. Soon, Gordon thought, Cape Cod will have an alternative to that dirty old eyesore.

Energy was on the minds of other Cape Codders as well. Senior scientist George Woodwell, founder of the Woods Hole Research Center, traveled to Washington's National Press Club to charge that the Bush administration's energy policy was nothing but drill, drill, drill. Ocean drilling, Woodwell said, was the last thing America needed, given the realities of global warming. Give emerging clean-energy technologies a try, he suggested.

Another environmental gadfly, Brian Braginton-Smith, uncredentialed but earnest, agreed. A well-known and much-liked local character who wrote letters to the editor and whose father owned a popular local hangout, Braginton-Smith dreamed of an "ocean ranch" that would farm both fish and wind. During the summer of 1999, he had gone to Boston to try to talk to Vice President Al Gore. Gore had kindly given him a few minutes, then directed him to the U.S. Department of Energy.

Boston Globe environmental writer Scott Allen picked up the story. Cute idea, doesn't have a prayer, he thought. Still, the concept was innovative. Allen wrote a story, which ran under the headline "Git along, li'l pogy: Plan has ocean cowboys raising food, wind energy." The journalist described Braginton-Smith's dream of boosting Cape Cod's prosperity by building a Nantucket Sound platform from which "ocean cowboys" would farm fish in pens and maintain roughly fifty electricity-generating wind turbines. Offshore wind turbines sounded to Allen new and exciting, if perhaps a bit outlandish.

The journalist's story along with a huge color photo ran on the front of the city news section, with a jump to an inside page. It was nearly impossible to miss. Among those who noticed were energy entrepreneurs Peter Gish and Brian Caffyn, just returned from building several land-based wind projects in Italy. They called Braginton-Smith to talk about a potential partnership.

Braginton-Smith then followed through on Gore's suggestion to contact the Department of Energy (DoE). He met with the DoE's Albert Benson, an older man nearing retirement who enjoyed mentoring and whose extensive and impressive résumé included a twenty-four-year stint with Mobil involving project development and financing. Soured on fossil fuels for a long list of reasons, which included geopolitics, air pollution, and the inevitable end of oil and gas supplies, Benson immediately saw the possibilities in Braginton-Smith's idea. An engineer with a degree in geology who served with the U.S. Army Corps of Engineers in both Vietnam and Korea in the late 1960s, Benson was an eminently practical man with solid experience in transforming the ideas of creative thinkers into real-world technology.

Having followed the development of wind technology, he knew the industry had recently come of age. Increasingly sophisticated engineering coupled with improvements in materials had brought the cost of a wind-generated kilowatt-hour down from thirty cents to ten cents over the past twenty years and the downward slide was continuing. During the same time period, the cost of a kilowatt-hour of electricity produced by fossil fuels had doubled and would likely continue to rise. Including a large amount of wind—which obviously had no fuel costs—in New England's electric generation portfolio mix would provide an excellent hedge on those escalating fossil-fuel prices.

Drop the fish farm idea, Benson advised. Focus solely on wind power. For New England, wind was the way to go. Then Benson suggested a bold concept: an offshore project of about 500 megawatts. A project that large, Benson said, would be a significant contribution to New England's energy supply and would be large enough to attract serious investors.

As electricity consumption per household continued to rise, Benson knew that communities would need to look at whatever indigenous resources they had. Cape Cod obviously had no fossil fuels, and the sun often hid for days at a time. One thing the resort area did have was wind. The wind blew off the ocean almost constantly, sometimes gently, sometimes viciously. Clearly that was the resource the region should exploit.

Benson had once overseen the management of Mobil's largest natural-gas-production field but had chosen to leave the company after a 1988 in-house study projected future severe natural-gas shortages. Time to get out of the fossil-fuel world, he thought. New ideas were needed. He worked temporarily for Massachusetts as an energy analyst, but decided he could achieve more by working with the federal government.

Mulling over the concept of a wind farm in wind-rich Nantucket Sound, Benson grew ever more enthusiastic. In Europe, wind was already substantial. Denmark, in fact, produced roughly 20 percent of its power from wind turbines. European governments had smoothly integrated the clean technology into their fossil-fuel-dominated electrical grids, and Benson had read about successful small-scale offshore wind projects in northern Europe.

The boldness of a 500-megawatt project excited Benson, who began to research how this new concept would proceed through the regulatory system. He produced a white paper saying that, under existing regulations, issuing project permits would be the responsibility of the Corps of Engineers, which was granted authority under the 1899 Rivers and Harbors Act. The Minerals Management Service of the federal Department of the Interior oversaw offshore oil and gas drilling, but when Benson called to ask what role that agency would play in permitting a wind farm, officials laughed. "We don't do anything but extractive stuff," they said.

He made one basic assumption: that the project would be considered based on societal need. No problem there, he'd thought. The case for societal need seemed obvious. All you had to do was look at Cape Cod's air pollution, among the worst in New England, or at the region's overdependence on natural gas. More than 40 percent of New England's electrical generation depended upon that soon-to-be-scarce commodity. New Englanders had to change, and change soon, Benson believed. Otherwise, it might be too late and the region might experience rolling blackouts. Wind could fill that gap, but, as with any new technology, the engineers who keep New England's lightbulbs glowing would need time to go through a learning curve.

While Benson and Braginton-Smith met, Gish and Caffyn were talking rather extravagantly about as many as 700 turbines providing as much as 2,400 megawatts of electricity. The duo wanted to build their project farther out to sea in an area known as Nantucket Shoals, several miles south of Nantucket Island. The site was impractical, distant from any adequate landfall connection and very vulnerable to the rough Atlantic storms. Many industry experts thought the proposal far-fetched, at least for the near future.

Then Jim Gordon showed up. He too had read Scott Allen's *Boston Globe* feature story. He too had started dreaming about a wind farm in Nantucket Sound. Gordon proposed that he and his engineering team, Braginton-Smith, Gish, and Caffyn team up. The Nantucket Shoals idea was shelved. The group budgeted $5 million in development costs, which in early 2001 seemed more than they would need.

Braginton-Smith began visiting various Cape Cod organizations to talk about the project that would soon become known as Cape Wind. Among those he visited was Chamber of Commerce Executive Director John O'Brien, who saw himself as an important player in the small-town political scene. O'Brien gave him the nod. He visited the daily newspaper, whose editors gave him the nod. He also visited Spyro Mitrokostas, the executive director of the Cape Cod Technology Council. Mitrokostas, forty-two, a London School of Economics grad and former Michael Dukakis presidential-campaign aide, thought the proposal had the potential to provide much-needed industry for Cape Cod, as well as steady, year-round, well-paid, high-tech employment. Far from seeing the sleek, modern turbines as industrial blight, he hoped the project would be *very* visible from the shoreline. Maybe it would encourage his children to become engineers.

"If this is Jim Gordon's project, why doesn't Gordon come and tell us about it?" Mitrokostas asked Braginton-Smith.

When Gordon, at six foot one and 165 pounds, walked into the room, Mitrokostas was impressed. He saw a man who had to be taken seriously. Gordon's closely cropped hair, athletic appearance, and neat but basic clothing belied his self-made wealth. The guy's substantive, thought Mitrokostas, who had expected someone a bit more flaky looking. He had little patience for environmentalists or aging hippies, but Gordon had a strong presence and spoke clearly, directly, and sometimes even eloquently.

Gordon knew the energy business and was obviously a no-nonsense type, but what was most striking was his intense earnestness. He concluded his presentation by asking for the sale, which also impressed Mitrokostas.

"I'd like your support," Gordon said.

"I'll help," Mitrokostas answered.

Then he warned the developer to be cautious.

"Only two or three hundred people run the Cape," he told the Boston-based entrepreneur. "If you don't have them on your side, forget it. If Ted doesn't like this, you're going to have a problem," he warned, alluding to the state's powerful senior senator.

On the other hand, he added, you might not have trouble with the crafty old senator. Kennedy seemed to have boxed himself in by spearheading the fight to keep the Arctic National Wildlife Refuge closed to drilling and by opposing drilling in North Atlantic waters. Given those well-known positions, to keep from being called a hypocrite Kennedy might *have* to support the wind-farm proposal.

As Gordon and his team continued to meet with Cape Codders, their belief that the project would be welcomed continued to grow. Only one politician seemed hesitant. Matt Patrick, the state representative interested in energy, had a run-in with Gordon and his lobbyist Neal Costello. The duo wanted money from the state's renewable-energy fund to pay for some of the wind farm's preliminary research, but Patrick refused. He had played an important role in getting the legislation passed that created the fund and believed it should be used for individual homeowners and for small public projects, rather than for private projects meant to turn a profit for investors. Gordon backed down.

The team moved into midsummer feeling pleased. Braginton-Smith reported that business leaders, local environmentalists, and the *Cape Cod Times* editorial board, including the conservative young publisher Peter Meyer of Osterville and his editor, Cliff Schechtman, liked the proposal. At the end of June, the *Times* had a small story that began by asking if project backers were serious. *Cape Cod Times* business writer Jim Kinsella talked with officials from several agencies, including the U.S. Army

Corps of Engineers. They seemed more curious than critical. Exactly what the regulatory process would entail remained murky. Nothing like this had ever been done in U.S. waters.

Using the wind to turn on a lightbulb is an old idea, as old as the lightbulb itself. In July 1859—the year that the teenaged Thomas Edison ran away from home to sell candy on a Midwestern train and the year that the middle-aged Colonel Edwin L. Drake would finally succeed in drilling the first commercial oil well in northwestern Pennsylvania—Moses G. Farmer of Salem, Massachusetts, delighted his young and rather dour New England wife, Hannah, by wiring up what seems to have been the world's first domestic lightbulb to a primitive battery to light up her parlor. Even the editor of the local Salem newspaper was said to be enchanted by the soft, bright light on Hannah's mantelpiece. Hannah was thrilled. All that month, as people from as far away as Boston came to troop through her parlor, the young wife glowed almost as brightly as her lightbulb.

Electricity was then the coming Big Thing, as exciting and modern as traveling to Mars might seem now. Sadly, though, Farmer's expensive platinum ribbon quickly burned out and the battery drained. Batteries in those days were simple, primitive, and very, very expensive. Invented early in the century in Europe, battery technology had developed incrementally in order to provide power to the fledgling telegraph industry, but the science of electricity as we know it today was nearly nonexistent. With the proof of the existence of the electron still decades away, the electrical force appeared ethereal and otherworldly. For the time being, playing with electricity was little more than an expensive hobby.

Hannah's husband, a brilliant and gentle man and earnestly religious, was motivated more by social morals than by money. As a youngster just off the farm, he invented America's first electric vehicle. It was

1847, the year of Thomas Edison's birth. Farmer and his brother took their creation—two carts hooked together, one with seats and a motor, and the other filled with a very primitive battery—around to village greens and gave rides for small fees. The young men never patented their carts. Indeed, Moses seems to have felt that patents were somehow immoral or antisocial.

Genius that he was, the good-hearted New Englander remained poor. He believed in developing technology for the common good, often refusing to take out patents even when advised to do so. For a while, he worked for several telegraph companies, but he was more interested in inventing technology than in being part of the company that operated what had been invented. He was in the wrong place at the right time. The old-fashioned Bostonians and Salemites could see very little use for an electric lightbulb. New England wealth was shipping wealth, based on that historically proven technology, the sailboat. A few men had invested in mills to make cloth and shoes—once the already developed technology was pirated from British mills—but in general the region's old burghers were not prone to novelty. Wind power was good enough for ships, water power was good enough for mills, and gas lighting was fine for home lighting. Why invest in an idea that might not come to fruition?

On the other hand, New York City, where massive amounts of capital were available for creative inventors, was a dream weaver's paradise. Farmer had failed to find financial backing in New England—and so did the inimitable Thomas Edison, a generation younger than Farmer and a telegraph expert who had come to Boston after the Civil War from the Midwest. When Boston turned down young Edison's ideas, he headed to New York City where enthusiasm for America's Industrial Age was revving into full gear. Edison's creative prowess and sheer dogged determination quickly grabbed the attention of J. P. Morgan and other up-and-coming moneymen. Morgan loved electricity. Unlike Boston's burghers, he understood its promise. As a youngster, Morgan had watched as his father, Junius Spencer Morgan, risked both money and personal reputation to finance the trans-Atlantic cable. After years of painful perseverance—the cable kept breaking—the project succeeded. The lesson in patience during the development phase of a technology was not lost on the young Morgan. Edison got the money he needed. In return, J.P. ultimately got himself an important new corporation—General Electric, formed in the last decade of the 1800s.

Yet even as they invented electrical products from lightbulbs to manufacturing equipment, the "electricians" of the age, Farmer, Edison, and so many others, struggled over a seemingly insurmountable handicap: the lack of affordable electrical power. Interestingly, the inventors knew early on that wind might be that source. Farmer wrote letters about the wind as early as the 1850s, trying to figure out how to cope with the annoying truth that wind didn't necessarily blow precisely when a homeowner wanted to turn on the lights.

In Cleveland in 1887, Charles Brush, a man who, like Edison and Farmer, seems to have been instinctively compelled to invent, found a solution. He built the world's first wind-powered generating system, a cumbersome thing that must have startled his wealthy neighbors. The machine had 144 blades and a 56-foot-diameter wheel. The electricity fed into a massive bank of more than 400 batteries in the basement of his home. The batteries in turn powered more than 350 house lights and several electric motors. The system worked well, in that it was comparatively dependable, but it was certainly not cheap.

Twenty years later, Brush's technology had matured enough that it had become affordable to the general public. Prairie farmers began buying small wind systems that could power their farm machines by day and provide electric light during long country evenings. Those systems continued to improve, but they were abandoned when large power plants became common and power became available in the cities at irresistibly cheap rates. The countryside, however, remained largely without electric power until the 1930s. Transmission and distribution wires were expensive. Power companies were happy to connect a large number of city residences to their central power source because of economies of scale. But there was little profit in running wires for several miles down a country lane to connect to a small home that might have only two or three lightbulbs and perhaps a radio.

During the Depression, Franklin D. Roosevelt created the Rural Electrification Administration, which ran wires to even the most isolated farmsteads. Roosevelt's gift was a miracle. When country folk saw their first electric lightbulb, perhaps in their children's school, some fell on their knees and cried. The lives of these long-isolated people changed in an instant. They had bright light to read by at night. Women had electric sewing machines, electric irons, and electric washing machines. Men had electricity-powered machines to help with the farmwork. The family radio brought them news of the world beyond their own mountain hollows or lonesome prairie farmhouses.

Wind technology seemed a thing of the past. Why bother, when you could buy central power for a few cents a kilowatt-hour? But in the 1970s, global politics coupled with gas lines, price jumps, and a consequent economic downturn, caused some to look again at the free energy available from wind. Jimmy Carter, wearing a sweater to ward off the chill, created legislation that encouraged the development of improved wind technology.

This legislation did result in the rebirth of the industry, but it also encouraged an excessive amount of financial hype. One tax incentive on the books allowed investors to get tax benefits by buying into the development of a wind-energy project—even if the project never produced any power. Hucksters plied their trade, vastly overstating the capabilities of the early-stage technology. Money flowed into the industry and, particularly in California, hundreds of wind turbines suddenly appeared. Some of those turbines worked. Many, however, produced little if any power before they broke down. And since investors received their tax breaks whether or not a turbine performed well, the useless machines were left to rust, with broken and twisted blades strewn across the ground.

None of this is unusual in the early days of the development of any technology. Ben Franklin's famous stove, invented so lovingly by America's first scientific genius, didn't actually work. When Franklin lit the stove, his family had to leave the house, which had suddenly become filled with smoke. It took another twenty years of development by other men before the stove, which Franklin never patented, could be used. Similarly, it took about twenty years, roughly one generation, before the modern wind industry came of age. The fast-talkers and shysters slowly disappeared.

The unsuccessful start-ups fell by the wayside. After the shakeout, a few serious companies emerged. In the United States, Zond Wind of California, successful in some of its early developments, was bought in 1997 by none other than Enron. Enron's wind division was bought, in turn, by none other than J. P. Morgan's now ancient behemoth General Electric, in 2002, under a court-supervised bankruptcy auction. GE's entry into the field was boisterously celebrated by the wind industry worldwide, which claimed that the grand old company's interest proved that wind as a technology had finally come of age.

By the turn of the century the company to watch was Vestas in Denmark, which had invested in serious engineering and had developed a reputation for manufacturing dependable machines that operated smoothly under all sorts of conditions. Wind's future looked golden, save for one problem. In Denmark, at least, the countryside had few remaining windy sites where turbines could be placed. Moreover, while the Danish people preferred wind-powered electricity to nuclear (they had actually voted after Chernobyl to forbid nuclear power within their boundaries), people living near older turbines found the noise annoying. Repowering—replacing older less efficient turbines with newer, more efficient, and quieter models—helped, but a broader new strategy was needed.

Europeans looked to the oceans. The offshore oil-and-gas drilling industry had honed ocean-based construction practices to an art. Moreover, wind resources over the water are far superior to those on land. With no mountains, forests, or buildings to break the flow of energy, ocean winds were usually stronger and steadier than inland winds. At least one European government, Sweden, was already experimenting with wind turbines in shallow ocean waters. Denmark had placed several wind turbines on a breakwater.

In 1991, near the small Danish town of Vindeby, which means "windy place," eleven turbines began turning in a shallow bight, followed by another group of ten in 1995 off the village of Tuno Knob. Other countries moved forward. In 1992, the Netherlands built a four-turbine project and in 1996, a nineteen-turbine one. By the late 1990s, offshore wind was the coming thing in Europe, raising hopes for a bright future for industry investors. Sweden and the United Kingdom chimed in with five-turbine and two-turbine projects, respectively, and the talk at all the conferences was of megaprojects with bigger and bigger turbines.

Offshore wind seemed to provide so many answers. First and foremost, of course, was that the new technology eliminated the need for fossil fuels. By translating the energy of the wind, essentially solar energy, directly into electricity, all the complexity of finding, mining, transporting, and burning fossil fuels would be eliminated. Once the technology was perfected, it was bound to be cheaper than fossil-fuel technologies, particularly if you included all those external costs. Attractively, offshore projects could be built near some of the world's largest cities, like London or New York, cities that consumed huge amounts of power, which often had to be transported from distant sites. Generating electricity nearby would vastly improve energy efficiency.

Erecting wind turbines in the water would also allow the individual machines to be larger. Land-based turbines have size limitations because of the difficulty of transporting the machines to the sites. Transporting blades more than 100 feet long on vehicular roadways was a serious deterrent, but such blades could easily travel by barge. In the early 1990s, a two-megawatt turbine was considered massive. By the end of the decade, Germany was bragging about developing a ten-megawatt offshore turbine. When asked how big offshore turbines might become, a Danish wind-turbine engineer, eyes aglow, responded that no one really knew. Visions of almost-free energy seemed to dance in his head.

The talk was heated and the words were sweet: If the wind-turbine industry could pull it off, it would be like getting something for nothing. That is, massive amounts of electricity could be produced with no pollution or fuel costs. If turbines could be built that could survive the cruel saltwater environment, wind would become the goose that laid the golden egg. Fuel would be free, energy would be clean, and a lot of money would be made.

Over on the other side of the Atlantic, Jim Gordon took notice.

If you stand at the tip of the town of Hull, Boston's high-rise buildings sometimes look as if they're next door. Yet driving to that tip, called Point Allerton, can easily take more than an hour in the nightmarish traffic of Greater Boston. Although the town and the city are close neighbors as the seagull flies across seven miles of Boston Harbor, to drive to the tip of Hull from the city, you have to head southeast out of Boston and then continue along the road, heading back northeast, almost completing a circle. The town is a very skinny peninsula, a series of islands, stretching north like a giant breakwater that guards Boston Harbor.

The drive to Hull is scenic. The ocean is almost always visible, and the peninsula's little humps of land, covered with late-nineteenth- and early-twentieth-century summer homes can look like the backs of whales. Hull was once a favorite of the Kennedy clan; Joseph P.

Kennedy Jr., older brother of the slain president, was born there in July 1915.

Jutting out into the ocean so precariously, Hull can also be a dangerous place. The Atlantic frequently wreaks havoc on the town, battering its seawalls and some of its houses, especially in winter nor'easters. After each big coastal storm, television crews routinely head there, knowing that a Hull resident can be counted on to appear on camera lamenting the hardship done to his sea-battered house. In storms, Hull is one of the Eastern Seaboard's prime television locales.

Perfect place for a wind farm, thought Malcolm Brown, a retired philosophy professor with more than a little of the 1960s left in him. Brown had built some small-scale renewable-energy projects in New York State before moving to Hull. His dream of wind power turned out to have

Malcolm Brown of Hull inspired the effort to purchase Hull's two new wind turbines.

historical precedent: The tip of Hull was called Windmill Point in the early 1800s. More recently, in 1985, during the early days of the modern wind-energy revolution, a very small, 40-kilowatt turbine placed next to the high school, almost at the end of the point, began generating electricity. The turbine, small as it was, reduced the school's electricity bill by 28 percent, according to a state report.

To Brown, that meant that the wind resource must have been fantastic. By 1997, the turbine was no longer functioning, due mostly, Brown thought, to the school's erratic maintenance. Brown and others formed a group called CARE (Citizens for Alternative Renewable Energy) and

John MacLeod of the Hull Light Department helped erect the first two modern wind turbines in New England, ultimately saving his rate payers hundreds of thousands of dollars.

approached the head of Hull's municipal light department, John MacLeod, a huge man who looks like Mr. Clean.

MacLeod managed the town's century-old light department. The idea of putting up a modern turbine appealed to him from the outset, not because he was an environmentalist, but because he was a dyed-in-the-wool American patriot. A Hummer owner, MacLeod had nothing against fossil fuels, but he did oppose dependence on other nations for the energy that lit Hull's residents' homes. Moreover, he knew a good financial opportunity when he saw one. By owning and operating its own electrical generation facility, the town could eliminate a long chain of middlemen, all of whom wanted a cut from townspeople's bills.

Brown and MacLeod didn't have a lot in common politically—Brown was pretty left, MacLeod right—but they found common ground when it came to energy. The vision of putting up a turbine on the tip of the windy town, a turbine whose glistening white blades would declare his town's independence from foreign oil, made MacLeod eager to get going.

But what, exactly, was the town getting itself into? Did these machines actually work? At about a million dollars a megawatt, were they dependable? Or was news of European success a marketer's hyperbole? Unaware of the ambitious project proposed for Nantucket Sound, the Hull group wanted to be first in New England to erect one of the new generation of efficient turbines, but they didn't want to end up embarrassed. When the light board learned about a wind-technology conference slated for late July 2001 in Albany, New York, they sent Brown fishing.

They armed him with kippered herring. A notice at the conference appeared on a bulletin board: Come one, come all to a party in Malcolm Brown's hotel room. Food provided, especially food for hungry Danes. Brown hoped to attract hungry and thirsty European turbine engineers and project planners and operators, so he could talk them up and find out as much as possible about the real-world functioning of these huge twenty-first-century machines.

They came. They ate. They drank. They talked. Among those talking was Peter Gish, of the team of Gish and Caffyn, who had made so much money building turbine projects in Italy. Gish hyped the new turbines in general and talked to Brown a lot about Vestas turbines. He also talked about the semisecret wind-energy project proposed for Nantucket Sound. In fact, the project was the buzz of the conference, thrilling wind aficionados like Brown.

A few days later, on July 28, 2001, a page-one story appeared in the *Boston Globe* announcing that an energy company was proposing an offshore wind project far bigger than anything yet built, right in the middle of Nantucket Sound—a favorite summer resort area for America's movers and shakers as well as for the international jet-set crowd. "The landmark project could place Massachusetts at the forefront of wind-power development," readers learned. "While a project of this scope is unprecedented in the United States, Cape Wind Associates LLC has assembled a formidable team with a credible prospect of getting its project spinning." The article described Gordon as a longtime innovator in the energy field with an excellent record of development, as "one of the first to take advantage of the deregulation of the power market, building seven power plants throughout New England with the combined capacity close to the Seabrook nuclear power plant. Since selling those plants last year, he is armed with a substantial war chest."

Now all New England knew about Gordon's ambitions. To some who noticed the sequence of events, it seemed as though someone at the conference had called the newspaper with the hot tip. But the truth was that *Globe* business reporter Jeff Krasner, previously with the New England bureau of the *Wall Street Journal,* had had extensive dealings with Gordon while he was developing his gas-fired turbine generators. Since coming to the *Globe,* Krasner had called Gordon from time to time to touch base. As the concept of the wind farm began to develop, Krasner coaxed Gordon into giving him an interview. After the wind conference, as talk about the proposal continued to build, Gordon agreed to go on the record.

Krasner genuinely liked Jim Gordon. The reporter had found Gordon a refreshingly frank man, a straight shooter who did what he said he would do. In fact, thought Krasner, Gordon was one of the few straightforward people in the power business. While other energy-industry execs were surrounded by suits, lawyers, public-relations staff, buffers, and handlers, Gordon had always seemed open and unpretentious.

Ominously, though, Krasner's *Globe* story included a statement from longtime Cape Cod politician Eric Turkington, a state representative from Falmouth and a passionate sailor. "I think they're out of their minds," Turkington said. "This is one of the premier yachting areas in the world and they're going to turn it into an obstacle course? That's leaving aside the aesthetics of having these poles sticking up out of the water for miles."

Not a sailor himself, Gordon didn't immediately grasp the dark overtones of Turkington's complaint. Announcing the project in midsummer 2001 was premature. With few of the project's details finalized, Gordon unwittingly opened himself to an array of charges from those who immediately opposed his project.

When they first got wind of some opposition, the team wasn't that worried. "Well, we all enjoy a good fight," they said to each other. Media advisers warned Gordon of the perils of having the news in New England's largest newspaper before paying his dues to the local community paper, particularly given its editor, Cliff Schechtman, and its publisher, Peter Meyer, an Osterville homeowner.

Bowing to the inevitable, Gordon had his staff invite the paper for an interview. In the dog days of August 2001, *Cape Cod Times* business writer Jim Kinsella was once again assigned to the story. It was an eerily quiet, hot, and humid summer day. Even the car traffic seemed to have stopped as Kinsella drove over to Gordon's Yarmouth offices. The air was completely still. The idea of the wind farm in the ocean sounded so outlandish that Kinsella couldn't judge whether it had any basis in reality or whether it was just a dreamer's fantasy. What would it take to implement something this ambitious? Did Gordon have the right stuff? Was Cape Wind a real company, or just some wannabe's New Age fantasy?

When he walked in and looked around the Cape Wind offices, Kinsella was surprised. The place was neither showy nor stripped down. It evoked an air of efficient seriousness. Kinsella knew that Gordon had a chance to make some real money—or that he had a chance to fail miserably. Who knew? What did seem clear was that Gordon seemed to be motivated by two dominating impulses, not necessarily compatible: the desire to make money, and the desire to do something to save the world.

Odd, Kinsella thought, but maybe the guy's sincere. Stranger things have happened. After all, for people playing in his financial league, there are lots of easier ways to make a mint of money. For Gordon, the reporter mused, this could be the gamble of a lifetime. You put everything on a number, and if it hits, you're a zillionaire. But it's an awful lot of trouble to go through to make money. Maybe Gordon really believed his line about "clean and renewable energy." Kinsella decided to suspend judgment and just tell the story of what Gordon and friends hoped to accomplish. Let the chips fall where they may.

Kinsella wrote a basic and straightforward story, frank and, seemingly, unalarming. Readers learned that an imaginative businessman with a strong and credible record of developing small-scale energy projects in New England wanted to build a very large offshore wind farm, in a locale that was, to some, a very important place.

The day Kinsella's story ran, August 9, 2001, was a landmark day for New England's electrical grid. The eerie atmosphere Kinsella had noticed the day before had turned even more ominously hot. In the midst of a days-long heat wave, ultimately broken by a late-afternoon thunderstorm, the grid was operating at full tilt, trying to meet a record demand of 25,158 megawatts. In Boston, consumers, some of whom had seen their rates jump by as much as 70 percent over the past nine months, experienced blackouts and brownouts.

Air quality was terrible. As often occurs on such days, a yellow-orange tinge hung over Cape Cod. Given the circumstances, some readers might have thought that Kinsella's story was both fitting and timely. In many places, reaction to the announcement of such a bold, innovative, and provocative idea might be immediate. But not on Cape Cod in mid-August. While Boston struggled to keep its lights on, Cape Codders were soaking up the last days of summer, gleaning the last harvest of sunlight and dollars before the long, cold winter to come. Very little was said—publicly, at least.

Surprised, Gordon hoped for relatively smooth sailing, but out on the yachts on Nantucket Sound, a storm was quietly brewing.

September 11, 2001, dawned clear and bright on Cape Cod, with temperatures warm enough that hordes of tourists remained to soak up the last of a seashore summer. The roads remained crowded. Owners of private planes and helicopters were readying for a day of filling the local airways, of ferrying the wealthy to and from their resort destinations.

Members of Osterville's Oyster Harbors Club, favored by families like the Mellons and the DuPonts, were breakfasting, giving orders to their household staffs and planning their outings to their little island's private golf course, their beach club, or their yachts. Matt Patrick was driving his hybrid car over a Cape Cod Canal bridge to Boston to work. Spyro Mitrokostas, the Cape Cod Technology Council director, was sitting in his office with colleagues, getting ready for a meeting. He had the television on to check the stock-market news. Jim Gordon was in his main office, at 75 Arlington Street, Boston, strategizing with his staff. It was business as usual, on another lovely late-summer day. One way or another, they all heard the same news at just about the same time.

At 8:46 A.M., American Airlines flight 11 crashed into the World Trade Center's North Tower, and at 9:03 A.M., United Airlines flight 175 crashed into the South Tower.

CHAPTER TWO

The Power Elite and Their Entourage

Democracy is the fig leaf of elitism.

Florence King

Old Joe Kennedy, paterfamilias and hard-edged kingmaker, had long ago advised his sons never, under any circumstances, to become involved in local politics. Stick to general issues, he advised the Kennedy boys, and don't allow yourself to get drawn into the morass of everyday problems. Keep above the fray. Publicly, the Kennedy clan has abided by that advice. They are frequently seen all over Cape Cod, on their boats and bicycles, in bars and restaurants, attending concerts or just shopping. And certainly the Kennedy clan is big business in Hyannis, with a few wags claiming the village had become a wholly owned subsidiary of its royal family.

But Kennedys almost never make public statements about local issues. For their part, local papers rarely write anything of consequence about the family, save for the occasional short news story when a

Kennedy ends up in court. It is as if the Kennedys exist in a separate sphere; they are *on* Cape Cod, but not *of* Cape Cod.

Behind the scenes, though, a great deal goes on. The younger generation is much given to inviting people over to the compound for drinks, a quick bite to eat (if there's any food), to go swimming, or just to hang out. Senator Ted and his young wife, Vicki, give "off-the-record" lawn parties for local journalists, very popular with *Cape Cod Times* editors. And the senator himself is quite prone to picking up the phone and making a few calls. For small-town politicians, getting a phone call from U.S. Senator Edward M. Kennedy is like being summoned by the king. More often than not, the senator will get what he wants. On Cape Cod, and particularly in Hyannis, Ted Kennedy is not someone you want to make unhappy.

Jim Gordon and company had intended to fly to Washington on September 12 for a much-desired and long-awaited meeting with the senator. Cape Wind executives had tried for months to set up a get-together, but all they'd received were arm's-length answers. Finally, though, they'd gotten a date. Gordon had prepared in depth for the meeting, looking forward eagerly to explaining all that Cape Wind could accomplish. If he had enough time, Gordon believed, he could convince Kennedy that his project could mark the beginning of a new energy era for Cape Cod and for New England.

There was, of course, no meeting. On the morning after the attacks, the nation remained frozen. Elected officials had far more pressing matters to discuss than wind energy. Gordon decided to wait several weeks, call again, and reschedule the meeting. He wasn't worried. After all, the senator's home base was Cape Cod. Getting together couldn't be all that difficult.

He was wrong. Phone calls went unreturned. Kennedy staffers refused to discuss the issue with Cape Wind, and no one could convince the senator that he at least owed Gordon a meeting. Gordon, not an innocent in these matters, sent in a campaign contribution and an RSVP for a fund-raiser. Kennedy staffers returned the check.

He's freezing me out, thought Gordon. He was soon to find out that Nantucket Sound and the Kennedy family had a long, much-prized association. In a sense, Nantucket Sound was accepted by many as the Kennedy clan's private pond.

On September 2, 1963, CBS's Walter Cronkite, who would decades later play a curious role in the seaside civil war, inaugurated America's first thirty-minute national news broadcast with an exclusive interview with President John F. Kennedy. The newsman and the president sat in wicker chairs placed on the sunburned grass of the Kennedy estate and talked via the television cameras to the nation. The glittering water of Nantucket Sound provided a fine background for this heady duo of power. It was the prestige heyday of Cape Cod's south shore, when Hyannis was mobbed with tourists, Hyannisport was mobbed with journalists, and Kennedy hangers-on milked their associations for all they were worth.

The Cronkite interview, one of the last times the president would see his beloved Nantucket Sound, wasn't the first time that political power had been wielded from these beaches. As early as the 1890s, high-level Republican politicians had summered here, drawn by the cooling breezes and Cape Cod's shirtsleeves informality. During the summer of 1896, Mark Hanna—future nemesis to Theodore Roosevelt, chair of the Republican National Committee and soon to be U.S. senator from Ohio—was said to have engineered William McKinley's presidential victory from a Hyannisport summerhouse. Ironically, when Hanna and his crowd arrived, they would have been living among a forest of derrick-like windmills that pumped fresh water for the summer homesteads.

Before the Civil War, Cape Cod was a little-visited backwater where only adventurous tourists dared to tread. Cape Codders felt themselves a people apart, going to Boston only when necessary to sell their fish or birds or to conduct a bit of legal business. Some left to make their fortunes elsewhere, but most were so insular that their version of the English language was barely understandable, even to other Massachusetts residents. Indeed, other New Englanders thought of the sand-spit as rather like a foreign nation. The Cape Codders themselves basked in their odd-man-out status.

After the Civil War, though, even insular Cape Cod changed as the Industrial Age, with its largely unregulated corporations, roared into being, spawning vast fortunes and much social misery. New piles of cash allowed the financial lords of the realm to send their families on summer-long seashore sojourns.

Cape Cod's western shoreline, made accessible by railroads, saw the first waves of this new affluence. Bostonians developed the commuting habit. During the hot months, the men could work in the city while their wives and families enjoyed the seashore. Many of the family summer homes built during that era remain today in the same families, some inherited through four or five generations via trusts. Often, for some odd reason, the furniture remains exactly where it was placed by the first, nineteenth-century generation. Books bought during the 1880s and 1890s are left to gather dust and grow mold pretty much where the great-grandparents put them down after reading them. Open up the medicine cabinet and you might find pharmacy products that long ago reached antique status. These strange summerhouse customs are thought to represent "tradition."

During that early era of resort development, much of the Cape's southern shore, without convenient railroad access, escaped the summerplace phenomenon. An 1880s Cape Cod atlas shows a distinct difference in development between locales with and without railroad lines. The village of Osterville, eventually to be filled with the fabulously wealthy, was only a hamlet with a few simple houses, home to a very few country families.

That was about to change. In the early 1870s, the Hyannis Land Company laid out a fairyland of streets across about a thousand beachfront acres and began marketing the Cape as "the Italy of Massachusetts" or "the Florida of New England." Why spend money on going to Europe when you could come to Cape Cod? The Hyannis Land Company failed after the panic of 1873. For a while, the lots remained largely unsold, but then visitors of a certain type began trickling in. They were searching for summer seashore adventures but were not drawn to the staid and low-key behavior of the Bostonians who summered on Buzzards Bay. Nor were they welcomed in places like Newport or Bar Harbor, where cash alone wasn't enough to get you on Mrs. Astor's list.

By the end of the century, nouveau-riche businessmen and politicians started arriving en masse on the Cape's south shore. Over on Yarmouth's Great Island, writes James O'Connell, Pittsburgh encamped, led by steel giant Andrew Carnegie, and his henchman, the ferociously antiunion Henry Clay Frick. Henry Phipps, another Carnegie partner who also had connections to the DuPont family, set up house with his brother, John. The depressed and eventually somewhat demented Henry Flagler, of John D. Rockefeller's Standard Oil Trust, arrived. Not all of them stayed long, but the flamboyant steel and oil men set the tone for this strip of coast.

By the early twentieth century, in published directories that fit handily into suit pockets or handbags and bore titles like *Dilatory Domiciles*, upper-class families proudly presented Cape Cod as their summer address. Others in their social class, or just wannabes, checked the summer-address register and followed the money. By the Roaring Twenties, the booklets showed Nantucket Sound to be the up-and-coming destination.

The southern shoreline in particular was famous as an anything-goes kind of place.

The new arrivals were distinct, culturally and financially, from old New England wealth. They were from St. Louis and Ohio and Pittsburgh, where the industrialists had formed a clubby group known for their frat-party closeness and crass behavior. The Pittsburgh crowd in particular were famous for running a martini flag up the flagpole to let their neighbors know they were serving drinks. Gin bottles were known to occasionally fly out of open second-story windows. Not attracted by such habits, New Englanders thought of the new-money folks as immigrants from the Wild West, from somewhere beyond the protective border of the Berkshire Mountains.

Nevertheless, they continued to come, particularly those pesky Pittsburghers. Among the earliest to arrive were the Thomas Bakewells, whose western Pennsylvania glassware had graced Dolly Madison's White House. The family of William G. Hawkins, long prominent in Steel City upper-class life, nestled in nearby. Pittsburgh businessman Walter Schleiter sent his family to East Bay Lodge, Osterville. They would be near the family of Armstrong Cork's Charles D. Armstrong, one of the village's earliest summer arrivals. The Thomas Evans family split their domiciles, one branch settling near the Bakewells, in Hyannisport, not surprising given that Evans money also came from Pittsburgh's Crescent Glass Works, and another in Wildwood Cottage in Wianno, that yachtsmen's enclave in the village of Osterville—so privileged that the collection of only a few streets had its own postal address, even though a place called Wianno didn't officially exist.

These people helped set the tone for the future Cape Cod—as a place not just of beauty, but also of privileged refuge, a place to escape the ravages of the very Industrial Age that had made them rich. They spoke of Cape Cod as a kind of Shangri-la, set apart from mainland America, where they could summer without the dirt, misery, and pollution of the smoke-belching, sky-darkening mills that had given them such financial freedom.

Bostonians of a certain type also came. Joseph P. Kennedy, whose fortune came from 1920s stock-market gambling and who didn't fit into the East Coast WASP crowd, also settled in Hyannisport. As the Cape Cod buzz spread, the trickle became a flood. Magazine articles and fictional stories about the region contained cloyingly sentimental writing, purveying hokum that was bought hook, line, and sinker. The hype spurred development. Marketing claims were fantastic. In 1926, Nantucket Sound's Craigville Beach became "the fourth best bathing beach in the world"—an interesting claim, given that most of the world's beaches had yet to be visited by developers, let alone catalogued.

Sailing was de rigueur. The Hyannisport Yacht Club held a regatta as early as 1909, long before old Joe Kennedy homesteaded the Sound. Along with the regattas came parties, roadsters, and alcohol. A local paper described life along the southern shoreline as a place where people did little "except to make whoopee and pursue dalliance." For most of America, the 1930s was a tough time, but in Hyannisport expensive cars carrying expensive people kept on arriving. Cape Cod's image as the American Riviera grew.

Oddly, the rich people built, of all things, imitation windmills, thought to be "quaint." Cape Cod once had many real windmills, which thrilled former Yale president Timothy Dwight, who was touring New England in the early 1800s. Dwight, apparently unfamiliar with such things, thought the technology wonderful. He described a windmill as "a pump furnished with vans or sails and turned by the wind." He rhapsodized over several pages. "The sight of these works excited in my mind a train of thought which others perhaps will pronounce romantic. I could not easily avoid thinking, however, that this business might one day prove the source of a mighty change in the face of this country."

A hundred years later, summer folk, eager to embrace as quaint what old Cape Codders had considered merely practical, bought up the old windmills. The antiques came to be so strongly connected to Cape Cod's national image that when one of the peninsula's favorite mills was bought and sent off to Henry Ford's tribute to American innovation—Greenfield Village, near Detroit—Cape Codders went into an uproar over the loss. Even the Oyster Harbors Club would choose an imitation windmill as its club logo.

"You're sure to fall in love with old Cape Cod," Patti Page crooned in the 1950s. By then, the yachting crowd was in full flower and sailboat races were thought to be the height of summer excitement.

In midsummer 2001, Rachel "Bunny" Mellon, empress of that same Oyster Harbors Club, made frantic calls to her Boston lawyers. You have to come down here and help us, pleaded Mrs. Mellon, a Listerine heiress then in her nineties. They want to put this wind-farm thing in front of my house.

At the entrance to the Oyster Harbors Club and housing development, an old-fashioned windmill and a No Trespassing sign stand guard at this gated elite community.

One lawyer answered that he genuinely liked the project.

You're a traitor to your class, cried this granddaughter of a Midwestern mouthwash manufacturer.

Bunny knew little about the wind farm, but she did have one fact correct: From her home on the south shore of Oyster Harbors, she would have an absolutely stunning view of Jim Gordon's project. Granted, on many days the turbines would be shrouded, either in mists or in air pollution. But there would always be those special days, those crystal clear days when visibility would be great. What if she wanted to have a garden party?

Bunny was a hard woman to ignore. The self-appointed queen mother during Jackie Kennedy's reign, she was so strongly connected to the Kennedy family that she ended up supervising the arrangements around Jackie's death as well.

Mrs. Mellon enjoyed much self-respect. Her grandfather was Jordan Wheat Lambert, who created and marketed Listerine via the Lambert Pharmacal company, based in St. Louis. His son, Gerard Barnes Lambert, was a major force in the American advertising industry. Before the Lambert father-and-son duo, bad breath was simply an unpleasant biological fact. But after their marketing campaign, it was a social faux pas of the first degree, something that would get you excluded from all the best parties. You could possibly become an involuntary recluse for life.

Over the years, the Lambert family made considerable social progress. From marketing bad breath they assumed responsibility for setting the standard for good taste in general. In 1922 Bunny's father started a phenomenally successful New York City advertising firm, the profits from which helped the family rise in social status. In St. Louis, the family helped finance Charles Lindberg's cross-Atlantic endeavor. Indeed, the aviator's plane took off from Lambert Field. Gerard Barnes Lambert devoted himself to sailing. By the 1930s, he was in the running for the America's Cup, a sailboat race requiring a fortune to compete in. Lambert was never a particularly successful competitive sailor, but he did instill in Bunny a strong respect for the kind of social life a yacht could buy.

Bunny Mellon's connection to the Kennedys derived from Jacqueline Bouvier Kennedy. Although she told a *New York Times* interviewer in 1969 that she and Jackie had been friends "long" before the Kennedys entered the White House, she seems in fact to have met Jackie in 1958 through Adele Douglass, wife of CIA deputy director Kingman Douglass and former vaudeville partner of Fred Astaire. Bunny, forty-six, and Jackie, twenty-eight, hit it off immediately. For one thing, Jackie *just loved* how Bunny had decorated her house. (So, apparently, did Bunny, who at one point hired an artist to paint pictures of Bunny's favorite rooms in all the houses she owned around the world. She then self-published a book of the paintings, which she handed out as gifts.)

When Jack and Jackie moved into the White House, the young First Lady felt very much at sea. Bunny promised to help. She was the right woman in the right place for that very job, having learned early that image is everything and that appearance can readily be turned into financial gain and social power. During Bunny's childhood, behaviorism was in its heyday. Psychologists claimed that the human mind was simple and mechanistic. Ring the bell, the dog salivates. Tell people long enough that halitosis is a social sin, they'll buy Listerine. So, naturally, when Bunny's father worried about his daughter's shyness, he had Bunny stand in front of a mirror and repeat: "You are wonderful, and beautiful, and the most glamorous young lady in the world, the prettiest young woman."

In 1948, Bunny left her first husband, Stacy Lloyd, to marry Paul Mellon, then said to be the world's richest man and scion of Andrew W. Mellon, U.S. Treasury secretary from 1921 to 1932, under whom three presidents were said to have served. Mellon money derived from a plenitude of fossil fuel. It came from Gulf Oil, Alcoa, and Koppers, a coal-industry enterprise—industries that resulted from the electrical revolution of the 1880s and 1890s—as well as from banking. As Treasury secretary, Andrew was best known for overseeing tax cuts for the rich and supporting a Wall Street free-for-all that ended in the Depression.

Andrew's son, Paul, was a bit of a lightweight, given to riding fast horses and writing mediocre poetry. Paul had steadfastly resisted pressure from his father to become an industrialist. He had a genuinely miserable poor-little-rich-boy childhood and was much at sea until he married Bunny.

Bunny, say people who care about these things, "brought out" Paul, socially speaking. By this it was meant that Paul became comfortable hobnobbing with the European upper classes and came to accept his lord-of-the-realm role as natural. In the mid-twentieth century, Bunny and Paul reigned supremely over the American socialite scene. Sally Bedell Smith's *Grace and Power* says it best: "Bunny and her billionaire husband, Paul Mellon . . . were the twentieth-century equivalent of Edith Wharton's van der Luydens, who 'stood above all of them' . . . ."

Paul was described as quietly imperial, while Bunny was often called simply imperious. It wasn't just that she and Paul liked to hang out with people like Queen Elizabeth II and the Duke of Edinburgh—in 1957 the British pair came to the Mellon Virginia hunting estate to drink tea and talk horses—it was that she seems to have set about intentionally engineering American culture, as her father had before her.

Almost as soon as Bunny was able to meet Jackie, the two women drew their husbands into a close relationship. Mellons, after all, gravitated to power, naturally assuming their right of association with presidents. Only nine days after Kennedy's inauguration, Jack and Jackie hosted a big White House reception that included the press. "Now the White House is a home!" wrote a newspaper editor, raving about the roaring fires in the fireplaces, the flowers everywhere, and the new State Dining Room bar where alcohol was served, much relieving reporters who had suffered severely during Eisenhower-era austerity. The flowers and other regalia, publicly presided over by Jackie, were in fact Bunny's creation.

Later in 1961, the Mellons gave an over-the-top "coming out" party for Eliza, Bunny's daughter by her earlier marriage. The extravaganza was so expensive that it brought cries of shame from the American public. "There had never before been such a party in the United States," wrote fashion journalist Dorothy McArdle. Jack had the good political sense to stay away from the royal extravaganza. Jackie did not. Among the many structures erected for the occasion was a lounge made to look like the Petit Trianon at Versailles, enjoyed so much by Marie Antoinette, wife of Louise XVI, built by Louis XV for his mistress, Madame de Pompadour. Bunny had created a fantasyland. The First Lady had no intention of missing the fun.

Bunny Mellon was the force behind Jackie's White House redecoration. Mrs. Mellon donated funds, supervised the garden redesigns, did some of the flower arranging, even donated or loaned art to the White House, either directly or via the National Gallery. Much of the elegance America saw during Jackie's televised presentation of the White House renovations was in fact Bunny's handiwork. When Jackie couldn't stay within her government allotment while redoing the family's White House quarters, Bunny provided cash rather than have the young wife risk the wrath of her father-in-law, who held the purse strings.

For Caroline, the First Lady wanted a play group, but whom do you invite for a just-us-gals kind of thing at the White House? The Mellons, of course. Bunny sent over her grown daughter Catherine, who had a child about Caroline's age with husband John Warner, the future Virginia senator.

Oyster Harbors, the gated community that held the Oyster Harbors Club, was a Kennedy refuge during the Camelot era. On August 13, 1961, for example, Berliners woke to a walled city. Jack learned the news as he and Jackie were preparing to sail along the Nantucket Sound coastline over to Oyster Harbors to lunch with the Mellons, a short sail of only several miles. Located on an island literally separated from the rest of Cape Cod by a drawbridge, a guard, and a gate, the club was essentially Bunny's project.

After the Second World War, the housing development on the island was nearly bankrupt, but Bunny loved the place. Paul agreed to finance the effort and eventually bought much of the island, turning it into a summer resort for Palm Beach types. A book about the Oyster Harbors Club, self-published by Oyster Harbors members, described the club Bunny believed she had created: "With its renowned golf course, country club, tennis courts and beach, and its more than 100 private estates, many owned by prominent families, it is one of the few retreats which seems to have assured its future and gracious life style for generations to come."

Bunny owned a long list of houses and apartments, but her Oyster Harbors home was special to her. Biographer David Koskoff wrote that when the Nantucket Sound sandbar in front of her home didn't meet her vision of a "Cape Cod beach," she had 2,000 tons of clean sand brought in from ten miles away and created her own dunes.

For years, Oyster Harbors had its own live-in state senator, Republican Edward C. Stone. Stone was a highly effective land conservationist who worked hard through the 1950s to lay the groundwork for protecting a number of important sites from commercial development, both on Cape Cod and elsewhere in the state. Without his hard work, the Cape Cod National Seashore might never have come into existence. The Kennedy family would eventually take credit for the seashore's creation, but Stone spent much of the 1950s meeting with local folks, trying to soothe their fears of loss of control to federal officials.

Stone was a dedicated and effective conservationist to whom the current generation owes a great deal. He also, however, was an elitist who knew who his supporters were and respected their needs. As early as 1950, the Massachusetts Audubon Society wrote Stone asking if the Mellons might want to donate some of the environmentally crucial barrier beach in front of their house. Stone wrote back with a polite but firm brush-off. Nantucket Sound beachfront was to remain in private hands.

He was not alone in that attitude. "I wish the hotel men were more alert to the problems which face the Cape," wrote C. D. Crawford, owner of The Pines, a Cotuit hotel, "with the double-barreling of the mid-Cape highway . . . a tremendous amount of traffic is going to be dumped onto the Lower Cape. Fortunately, it will by-pass Osterville and Cotuit. I do feel, personally, that the problem of public beaches is most important. We cannot maintain our private beaches unless we provide waterfront for the masses . . . ."

Another waterfront-property owner, a William R. Baker, wrote to Stone for assurance that none of the lands slated for public access would be near his property. Replied Stone: "The part of Barnstable County where you live will not be in any wise affected. Nobody has a greater interest to see you are protected than I do."

Soon after Bunny Mellon's frantic phone call, Barnstable town councilors began working themselves into a serious tizzy. There are always politics involved in the acceptance of any big idea, but in this case, the war-dance drums sounded long before most Cape Codders had any idea what was happening.

Barnstable town councilors Gary Blazis and Royden Richardson, local men who frequented the gatherings in Osterville and Wianno and Hyannisport, intended to nip Gordon's proposal in the bud. In early October, the duo blindsided the Cape Wind team by introducing a town council resolution claiming the wind farm would have a "devastating" effect on the Cape's boating, tourism, and fishing. In addition, the resolution claimed, the towers would be visible twenty miles away.

The absurd claims astonished people. The wind turbines certainly wouldn't normally be visible twenty miles away; Cape Cod tourists went to many places other than Nantucket Sound; and boaters sailed in plenty of places other than on Horseshoe Shoal. The project would be spread out over about twenty-six square miles of Nantucket Sound, with turbines laid out in rows about one-third to one-half of a mile apart. But the Sound spreads over hundreds of square miles, making the claim that Cape Wind would "destroy" the Sound outlandish.

Among those alarmed by the precipitous resolution was Matthew Patrick, the state representative. Worried about escalating energy prices and familiar with modern wind technology, Patrick wondered why the town council wanted to strike so preemptively.

At the town council meeting convened to discuss the resolution, Patrick rose to speak. "Let's not throw the baby out with the bathwater," he pleaded. He explained the significant clean-air benefits for Cape Cod and suggested that the project could reduce overall electricity costs for all New England power-grid users.

Other knowledgeable people also spoke. Charles Kleekamp, a retired engineer and clean-air advocate, introduced data to back up the claim that Cape Wind would significantly decrease sulfur oxides, nitric oxides, and carbon dioxide—pollutants that routinely escaped from the Cape's oil-fired electrical plant smokestacks. Cape Codders, Kleekamp said, had seen a large increase in the cost of electricity in recent months. The wind project, which would not be subject to fluctuating fuel costs, would dampen those increases.

Kathryn Kleekamp, Charles's wife and a retired medical microbiologist, talked about the severe health effects resulting from fossil-fuel-fired electrical generation. She had several times traveled to Washington at her own expense to give similar testimony at Capitol hearings on clean-air issues.

Town council president Richardson, suddenly impatient, harshly interrupted the scientist. At one point, he tried to cut her off, not an easy task. She concluded her short presentation: "I rather imagine there are far more people who don't have waterfront property, who don't boat or fish and yet would highly benefit from the positive health impacts of clean, locally available wind power," adding data about the increasing number of high-ozone days each summer on Cape Cod.

Then came several southern-shoreline marina owners. "I can't picture the customers we've been cultivating to come here and people that come in here and anchor off the yacht club" will want to see "all that machinery," warned Wayne Kurker, soon to become a major opponent.

Greg Egan of Crosby Yacht Yard, located near Oyster Harbors and once highly favored by Paul Mellon, said his customers "from literally all over the world" thought of "Cape Cod and Osterville as a place of natural beauty" that must not be jeopardized by the wind farm. These yachtsmen, Egan added, would have to sail right by the project as they piloted their boats from Newport to Osterville. They would not, Egan warned, be pleased by this unfair burden.

One of the odd things about this opposition was that Cape Wind had yet to make a formal public presentation about exactly what the company wanted to build. Few, if any, specifics existed. Feeling railroaded by the preemptive resolution, Gordon did not come to the council meeting, but he did send engineer Craig Olmsted, who was given a few minutes to present. Olmsted was hard-pressed to explain much, since the councilors were obviously unfamiliar with wind technology and knew next to nothing about the basics of electricity flow, or about the electrical grid, or even about Cape Cod's extant oil-fired plant.

Clearly frustrated and somewhat confused by the councilors' hostility, Olmsted was reduced to begging councilors to "consider very carefully all the facts."

An indignant Richardson, a sometime justice of the peace who led prayers during Osterville gatherings, interrupted him. "I would like to lay this question to rest right now," the council president intoned with considerable deliberation. "Everything this council does we consider carefully and we try to be as thorough as possible."

"And, also, fair," he said slowly, as an afterthought.

How can you be fair by opposing something you don't know anything about? Matt Patrick wondered. At one point, Patrick tried to speak up for allowing the wind-farm team more opportunity to reply to councilors' concerns. Richardson ignored him.

Patrick, a relative neophyte to power politics, learned a lot that day. He came from nearby Falmouth, a town very different from Barnstable. On Cape Cod, time is multidimensional. To drive from one town to another may only require twenty minutes by automobile, but you may well be traversing several centuries of cultural separation.

Falmouth had a long history of hosting several world-class scientific institutions, including the Marine Biological Laboratory, founded in 1888; the Woods Hole Oceanographic Institution, founded in 1931; and the Woods Hole Research Center, founded in 1985 by George Woodwell. These scientific powerhouses were all stuffed together on the tiny peninsula of Woods Hole, as different in character from Oyster Harbors as Princeton's Institute for Advanced Study is from the Palm Beach Bath and Tennis Club.

Watching the Barnstable council, Patrick began thinking to himself that the phrase "review process" must have a different meaning in

Hyannis than it did in Falmouth. There must be two different types of politicians, he thought: those interested in facts and informed decision making, and those who made their decisions based on other criteria.

Patrick had told councilors that the wind turbines, placed four to five miles out to sea, would be only the height of an outstretched thumb placed along the horizon line when seen from shore. An irritated-looking Councilor Gary Brown, close associate of Richardson and Blazis, returned to that idea. "I drive by Craigville Beach about three hundred times a year," he said, "and I don't want to have my thumb out the window blocking these as I drive by!"

Patrick was stumped. It was hard to know how to respond to such a remark. The location that the Cape Wind team had chosen, Olmsted had explained, was in federal waters. But as Councilor Brown continued, he made clear his own thoughts on that issue: "In Nantucket Sound there are no federal waters!"

Again, Patrick was stumped. Had Cape Cod seceded from the United States? Years ago, first in the early 1980s and then again in 1986, in *United States v. Maine*, the U.S. Supreme Court had decided clearly that the center of Nantucket Sound came under federal rather than state jurisdiction. State waters extend three miles out into the ocean, but there, jurisdiction stops. Though some federal legislation gives the state limited purview in waters farther offshore, the ultimate decision about use of those waters remains with the federal government.

Nonplussed wind-farm proponents tried to coax the councilors into considering the matter more thoroughly. Think of the possibilities here, they suggested. Cape Cod could become the leader in a totally new concept in electrical generation. But the councilors were having none of this kind of talk. This project has been "jammed down our throats," sputtered Councilor Blazis, so upset he could barely get his words out.

One of those who turned up to speak in favor of the project was none other than the Department of Energy's Albert Benson, who had continued to follow the wind project's progress. "September 11 changed a lot," he said. "You and your constituents have a chance to actually get involved in doing something that's beneficial to the national initiative." Here, said Benson, was a way for regular citizens to make a difference.

As Benson listened to the ravings of the councilors, he grew concerned. Facts didn't seem to matter. The councilors seemed immutable. Benson decided the Barnstable councilors were operating out of some pre-Copernican concept, some sort of belief that Cape Cod was the center of the world and that nothing else mattered. For a man who had spent much of his career working in Japan, Indonesia, Saudi Arabia, and elsewhere, he just didn't know what to make of Barnstable officials. But when he brought up the concept of patriotism, the councilors sat stone-faced.

Some project proponents believed that after that first outcry, Barnstable's elected officials would undergo a learning curve and their opinions would change. Gradually, however, they gave up that idea. In meeting after meeting, called by an ever-lengthening list of local, state, and federal agencies, the same hysterical exaggerations were made, until those first gatherings, so similar in rhetoric, merged into one long tedious nightmare for Jim Gordon.

There were a few calm voices. At that first Barnstable meeting, Councilor Richard Elrick, a Nantucket Sound ferryboat captain, let loose: "The use of wind power is of huge advantage to us," Elrick lectured his colleagues. "Unfortunately, this resolve was presented to us this evening with absolutely no facts. . . . It would have been more appropriate for the proponents of this resolve to have come forward with such a thing if they felt it was appropriate after we at least had a discussion!"

Gradually, project supporters began to emerge. At one public hearing in late December sat thirty-nine-year-old Matt Palmer. Palmer, a plant engineer, had been working at Gordon's former gas-fired power plant in Dighton, Massachusetts, for six years, ever since the plant itself was just a hole in the ground with some concrete and a few steel columns. When Gordon sold the power plant to another energy company, Palmer had stayed on. Gordon had asked Palmer to attend a public meeting and talk about Gordon's past record developing power plants. Palmer agreed, but, knowing nothing about wind power, wondered what his former boss was up to. He imagined that Gordon's turbines might be floating at sea on some sort of platform, tipping back and forth like buoys in Nantucket Sound. Bizarre, thought Palmer. Had Gordon gone off the deep end?

At the meeting, he listened as Olmsted explained that the turbines would be firmly anchored to the ocean bottom by one of several techniques already sharply honed by marine construction firms experienced in building ocean structures. Not so flaky after all, Palmer thought. He learned that turbines at sea could be larger than turbines on land, that the winds were steadier and therefore more productive, and that after ten years of research, European scientists had discovered no environmental drawbacks.

When Palmer did the math and figured out how much power Gordon's proposed project would generate, he was astounded. It was identical to the gas-fired power plant where he worked. And the wind farm would be producing power day after day, releasing plant managers from worrying about either the cost or transportation or acquisition of natural gas. Why in the world would we want to burn fuel to make that much electricity, Palmer wondered, when we can let the wind do it for us?

After that meeting Palmer began calling himself a "recovering fossil-fuel guy."

Gordon had walked directly into a trap. Politically naïve, he had never imagined the scope of resources that would be called into play as the

Cape Cod power elite closed ranks or how quickly they would close ranks. In response, he began working at the grassroots level, trying to contact as many community groups and town boards as possible, trying to create a groundswell of support. Often, his staffers were simply stiff-armed, either not allowed to speak at all, or ridiculed after being given only a few minutes to speak.

Surprisingly, he was permitted to address the Osterville Men's Club. If Osterville was one of the richest and therefore most influential of Cape Cod's villages, the Men's Club was its decision-making old-boys watering spot. In the rest of America, men's clubs might seem anachronistic, but in Osterville, club members take themselves quite seriously. Among those present that day was *Cape Cod Times* publisher Peter Meyer. A men's club seemed an odd place for a newspaper publisher. What's he doing here? Gordon wondered, not yet privy to the fact that the young Meyer owned a $1.2 million home on Osterville's Wianno Avenue.

Gordon was not allowed to address the Cape Cod Chamber of Commerce. When he asked to be invited, he was told instead to come to a small executive board meeting, at which he was allowed to talk for about ten minutes. Repeating his request to address the chamber as a whole, Chamber president John O'Brien put him off. We'll get back to you, O'Brien promised.

It never happened. A few days later, O'Brien and the Chamber issued a press release opposing the project and making a series of grossly false claims, among them that "in a deregulated electricity marketplace, New England will soon have an over-supply of power." The statement directly contradicted information provided by New England electrical-grid managers, who were increasingly worried by the region's dependence on natural gas.

Shortly after issuing that press release, O'Brien registered as a paid lobbyist. His son, a former state legislator, worked for Sithe Energies, which owned several New England fossil-fuel-powered electrical plants. If wind generation were to become common in the region, those plants might be financially affected. Many wondered whether Sithe was encouraging O'Brien to stop Cape Wind. O'Brien's lobbying firm, the Minehan Companies, represented Sithe.

Additionally, O'Brien became a lobbyist for the Alliance to Protect Nantucket Sound, which he had in fact helped organize. This lobbying connection was kept secret from the general public. Indeed, when asked, Alliance employee John Donelan denied O'Brien's lobbying position. And when *Cape Cod Times* columnist Francis Broadhurst tried to write about it, *Times* management told the writer that O'Brien's position as a paid lobbyist was "not relevant" and that he could not put the fact into print. When asked directly by a reporter about the lobbying relationship, O'Brien would respond: "How did you find that out?" When told that the documents he'd signed about his lobbying role were public record, O'Brien seemed surprised. "This is like a pimple on an elephant," he said.

As the fearmongering and the hysteria mounted on Cape Cod, New England's first modern, commercial-scale wind turbine began turning fifty miles to the north. Up in Hull, John MacLeod pushed the switch to begin official production at 2:45 P.M. on December 27, 2001. Malcolm Brown was there to take pictures and savor the moment. By year's end—four days later—the turbine had already produced about 20,000 kilowatt-hours. Four years later, the one turbine had produced well over 6 million kilowatt-hours, or 6,000 megawatt-hours. The higher-than-expected production, coupled with the town's conservative original financial estimates, meant that the turbine had nearly paid for itself within those four years, rather than the projected twelve.

CHAPTER THREE

The Electrical Priesthood

. . . that men shall say of succeeding plantations, the Lord make it like that of New England . . . .

John Winthrop, 1630

James Jackson Storrow was a phenomenon so rare as to be virtually nonexistent in late-nineteenth-century Boston: a Yankee Democrat. On his mother's side, he came from Boston Brahmin stock—Cabots, Lees, Higginsons, and the like. Commodore Oliver Hazard Perry—of "We have met the enemy and he is ours" fame—was a Storrow ancestor. But on his father's side, the bloodlines were a bit murky, if not downright suspicious. There was a tale about a Jamaica planter, captured during the Revolutionary War and brought to New England, who then married a New Hampshire Wentworth. To proper Bostonians, Storrow's pedigree was not entirely reliable.

Storrow seemed to enjoy the mystery. When the banker wanted to amuse himself, he told people that some of his ancestors came from the South, and then allowed them to decide for themselves precisely what "the South" meant, and whether this aberration might account for his unusual political views. He could, of course, afford to be mysterious. By the beginning of the twentieth century, James Jackson Storrow was often said to be Boston's most important citizen.

Storrow was a proponent of progressivism and the good-government movement then sweeping America, led by Theodore Roosevelt and many others. By most accounts Storrow was personally a kind and thoughtful man, albeit publicly rather austere. On the other hand, he was also a tough-minded banker and businessman who took his Puritan-endowed responsibilities seriously.

Storrow literally changed the face of Boston. He was a driving force behind the damming of the Charles River and the construction of the river's boat basin. His wealthy Beacon Street neighbors fought him every step of the way. They claimed his project would ruin their water views. A business and civic visionary, Storrow became General Motors chairman temporarily in the early 1900s in order to reorganize and reinvigorate the heavily indebted corporation. He was president of the Boston Chamber of Commerce and worked hard to solve the problems created by J. P. "I-owe-the-public-nothing" Morgan's monopolistic control of New England railroads, which many believed charged excessively high fees while providing excessively poor service.

Storrow and his wife had only one son, but both liked children. In particular, they were well known for helping the sons of immigrants get established in the often forbidding Boston business culture. Storrow helped some boys find jobs and provided others with money for college. He had playgrounds built for poor children on the banks of the Charles, revamped the corrupt Boston School Committee, and served for a while as president of the Boy Scouts.

Immigrants to Boston faced a set of unique problems. The city—called in 1858 "the hub of the solar system" by Oliver Wendell Holmes—considered itself the leading light of the nation, and, indeed, in those years the city was a genuine culture hearth for literature, philosophy, education, psychology, and other progressive ideas. Boston Puritans had a historical tradition of respect for the printed word, and all children were supposed to learn to read.

On the other hand, Boston Puritans presided over a rigid, closed society that insisted on strict conformity to a set of sometimes rather harsh rules. Individualism was firmly prohibited. Just because reading was a good thing, it didn't necessarily follow that freethinking was acceptable. In the 1600s, New England town fathers had the right to run people out of town when they wanted to—and they often did. They were not reluctant to use their powers. Quakers had their ears cut off. Baptists were tarred, feathered and sent to Providence.

After the Civil War, as the country changed, so did New England. Control began to break down. There was employment galore in New England's new factories, so much so that the region's supply of farm girls no longer sufficed. Waves of immigrants arrived, seeking work. After well over 200 years of relative peace, conformity, and isolation, change was in the air. Boston's Brahmins were less than certain that variety was the spice of life, but there seemed little they could do to rescue their once-homogeneous paradise.

Boston changed in ways that the old families deplored. Two sections in particular, the North End and the West End, were infamous in the minds of the Yankees. There, the historic buildings were evacuated by the old line and resettled by new families, who often could afford no more than one room. Those rough neighborhoods were avoided by most of the old crowd, but Storrow, an inveterate walker, was often seen there, strolling around, talking to children, visiting civic groups, and looking for ways to improve the lives of the city's poor. On the surface, Storrow appeared all Yankee—tall, taciturn, righteous. But he also possessed a genuine compassion for suffering people and he was prepared to act upon that emotion.

In the early twentieth century, Boston's fifty-acre West End was like New York's Lower East Side. Its 20,000 residents hailed from Ireland,

Italy, Greece, Lithuania, Russia, China, India, and more. Several hundred blacks lived there, as did many Jewish immigrants. The West End's many five-story tenements, lacking adequate cooking, sewage, or bathing facilities, were perfect petri dishes in which were cultured a long list of deadly diseases, from tuberculosis to alcoholism. After school, gangs of street kids, often with ethnic affiliations, ran wild. The kids were at a loss as to what to do with themselves, other than to hang out in the streets. The city of Boston provided West End kids with the nation's first public bathhouse, but men like Storrow understood that such facilities were only a small first step.

In 1903, at just about the time that Cape Cod's southern shoreline crowd was forming its many exclusive yachting clubs, a small group of West End teenage Jewish boys formed their own club, called first the Bootblack League, then the Excelsior Club, and then the ambitious Young Men's Excelsior Association. The club discussed ideas, performed Shakespeare, and organized escapes from the West End's misery. Hearing about the club, Storrow gave them a forty-volume complete works of Shakespeare. As thanks, the boys invited Storrow to a meeting. Then Storrow invited the boys to spend a Sunday at his huge estate in Lincoln, a town just outside Boston. A year later, Storrow funded a building for the club. Eventually renamed the West End Boys Club, it expanded to include boys from all the West End's ethnic groups and became one of the founding clubs of what are now the Boys and Girls Clubs of America, with 3,700 clubs nationwide and more than 4.4 million members.

Simultaneously, Storrow and department store owner Edward A. Filene founded the Boston City Club for the betterment of adults. The driving force was not exclusiveness, but inclusiveness. Intended as a "meeting place for opposites," the club was supposed to welcome a wide range of people and ideas. "May the time never arise when our friendliness will not be as generous and as cosmopolitan as our great city, and our open-mindedness and sympathy as broad as the human mind," Storrow wrote of the club's goals.

Because of his progressive interests, Storrow eventually found himself running against Ted Kennedy's grandfather, John "Honey Fitz" Fitzgerald, in Boston's 1909 mayoral race. Fitzgerald had started his political career as a North End ward heeler—essentially a tattletale to the ward boss. As a young man, it was Fitzgerald's job to rat on voters who did not mark their ballot for the proscribed candidates. While monitoring the ballot box, if the young ward heeler could tell a voter was choosing the wrong candidate, he raised his arm, bringing over enforcers who confronted the independent-minded voter, sometimes by depriving him of a job, sometimes by physically beating him.

By the time Fitzgerald became Boston's mayor, he had become connected to some highly visible financial scandals. Seeking to defeat him, Boston progressives put up Storrow to oppose him in a three-way race. Storrow seemed a good choice for the reformers. Attracted by his pedigree, old-line Republican Yankees supported him. The newer wealthy families knew him through his banking connections and the city's voting public knew him for his good civic works. He seemed a shoo-in.

The only strategy open to Honey Fitz was to attack Storrow's wealth. Fitzgerald repeated an endless litany of sound bites that pointed toward Storrow's social status, positioned the election as a battle between Harvard and the Boston slums, and used the campaign slogan Not For Sale, coupled with a picture of City Hall, implying that Storrow was trying to buy his way into government. The Fitzgerald campaign also circulated a picture of Fitzgerald with his wife and six children, with the slogan Manhood Against Money, presumably implying something about Storrow's virility. Whisper campaigns claimed that Storrow disliked various ethnic groups and intended to prevent the working man from improving his life.

Unable to overcome the lies, Honey Fitz's positioning of him as a "rich" Yankee, and the whisper campaigns, Storrow lost by a narrow 1,500 votes. Had there not been a third party, a Republican spoiler, who gleaned 1,600 votes, Storrow might have won despite Fitzgerald's dirty

Energy entrepreneur Jim Gordon.

campaign. Afterwards, apparently disappointed but not embittered, Storrow wrote to his friend Filene that "it was racial and class feeling, and the consequent unwillingness to consider any of the questions with an open mind, which defeated me."

Just as Ted Kennedy traces his lineage back to Grandfather Fitzgerald, Jim Gordon, in a sense, traces his lineage back to James J. Storrow. While Honey Fitz was wending his way through Boston ward politics, Gordon's paternal grandfather was emigrating from Russia to escape the relentless pogroms, which had so distressed President Theodore Roosevelt that he told a delegation of American Jews that he shared with them a deep feeling of sympathy over the slaughter. With a brother and several cousins already in the West End, it was there that

Gordon's grandfather settled, married, had three sons, and worked as a butcher in a meatpacking house until he was seriously injured by a falling meat hook. In those days before workman's compensation, insurance, or any other social safety nets, the family barely scraped by.

The West End Club helped take up the slack. Jim Gordon's father and two uncles both joined the club and found opportunities there not otherwise available to them. Among the most important and most memorable of those opportunities was the chance to spend summers at the West End Camp in Maine, founded by Storrow, a great outdoorsman. At the end of the nineteenth and the beginning of the twentieth centuries, the outdoor life was very much in vogue among the educated classes of New England. Among Boston's leading intellectual lights was the professor and psychologist G. Stanley Hall, whose 1883 publication *The Contents of Children's Minds* claimed that 91 percent of ghetto children had no idea what an elm tree was. Hall and the other horrified reformers, adherents of Thoreau and Emerson, believed that healthy New England children absolutely required the experience of New England nature.

Life at the Maine camp wasn't subsistence level, but neither was it luxurious. The boys lived in canvas pup tents, which could get quite chilly in a northern summer downpour. The camp ethos emphasized teamwork, competition, and physical prowess. On Friday nights, two boys were thrown into a boxing ring and the rest of the boys watched while noses were bloodied. Younger boys were often sent out alone in the woods at night by older boys, presumably to test their courage, and everyone pitched in to do all the basic chores of camp, from cooking to latrine maintenance. Early life in Maine's West End Camp seems to have been a bit like a cross between *Lord of the Flies* and Marine boot camp.

The atmosphere fit Jim Gordon superbly. As soon as he was old enough, Gordon's father trooped him down to the clubhouse on Blossom Street in the West End to sign up for the camp. In keeping with club tradition, boys began attending at age eight. Most attended for part of the summer, but Gordon went for the whole eight weeks. With the family living in a rented house in Newton Center, just outside Boston, his parents felt that the camp would both develop his self-confidence and keep him off the city streets for the summer. Gordon looked forward to joining his older brother, Michael, who had already attended for several summers.

Nicknamed "Toughie" by the age of five, Gordon thrived in the Maine woods, where his energy and determination could be expressed in a way that wouldn't have been permissible in Newton. He didn't cause serious trouble, but he was definitely an instigator. Camp counselors had to keep a close watch on this slight, wiry, and endlessly energetic boy with a very creative mind and an assertive character. For Gordon, even as a kid, the road less traveled was not an *option,* it was a *compulsion*. Counselors who remember him tell stories about flaming arrows shot through the air from cabin roofs, and about how thankful they were that no one was actually set ablaze.

At thirteen, his camp nickname became "Cool Hand Luke," because no matter how grueling the punishment Gordon seemed simply biologically incapable of calling it quits. One late afternoon, he and an older boy, a camp counselor who was on the Colby College basketball team, started a game of tetherball. Outmatched in age, weight, and height, most thirteen-year-olds would have enjoyed the fun, accepted an inevitable loss, and moved on. Not Gordon. Every time the older boy had almost wrapped the cord completely around the post, Gordon would somehow, at the last minute, jump up and use all his weight to hit the ball back, unwinding the rope. He couldn't win, but he was unwilling to lose. As word of the competition spread throughout the camp, more and more boys showed up to watch. After an hour and a half, the crowd had swelled and Gordon was in his glory, at the center of the action, trying to win an impossible battle. He never did give in. Eventually counselors called the game because of the dark and sent the boys off to the mess hall for dinner. Gordon never forgot the fun of being the underdog.

Friday-night fights were another time when all the camp gathered. Boxing was an important rite of passage for the boys, who sometimes volunteered to go into the ring and who were at other times volunteered by the counselors. Two boys who had been arguing during the week were likely to end up with gloves on. Gordon came from a middle-class suburb, but other kids at the camp came from some very tough neighborhoods. In one match, Gordon survived an experience in the ring with one kid who, years later, ended up being killed while committing an armed robbery.

The camp intentionally fostered a competitive atmosphere. During the final weeks of summer, the boys divided into two teams to compete for a series of honors. The annual competition, called Color War, was as serious to the kids as any pro sports battle. The teams even wore "uniforms"—blue T-shirts for one side, white for the other. Not all the competitions were athletic. At night the two teams competed in spelling bees, in writing songs and plays, or on a scavenger hunt. The mess hall was normally a noisy, rowdy place, but during the competition, no boy was allowed to talk during meals. Discipline was deemed essential and any boy who broke the silence lost points for his team. Gordon quickly became a team coach and brought his side to victory. The thrill of winning at team coaching was also a life lesson.

When not at camp or at school, Gordon worked in one of his father's two corner stores in working-class Allston, a Boston neighborhood. Beginning at the age of twelve, he did basic store management, lugging groceries up from the basement and putting them on shelves, ordering from vendors, and selling items to customers. He had a lot of responsibility for a preteen. Both stores were located in the midst of large apartment buildings with lots of families. Customers dropped in once or twice a day to pick up bread, milk, or a newspaper and to pass along gossip. When not visiting with customers or stocking shelves, he read one of the many papers the store sold.

Gordon loved it. The mid to late sixties was a glorious time to be a kid in Boston. Change permeated the culture. On one side of the two Allston stores were Harvard and MIT; on the other, Boston University and Northeastern University. During his high school years, after school he took a trolley to the store, turned on *the* cool radio station of the era, WBCN, and listened to Jimi Hendrix while he stocked shelves and talked to the families, stay-at-home housewives, local drunks, and hippies in their fringe jackets who dropped by.

The little stores were open 365 days a year, from 7 A.M. until midnight. Working there, Gordon learned that people who ran their own businesses did not work eight-hour days, did not take long vacations, and did not have much leisure. They worked hard and unpredictable hours, but they did have the opportunity to manage their own lives. Even as a teenager, it looked to him like a pretty good trade-off.

After high school, he attended college in Florida for a while, then returned home to complete his education at Boston University, from which he graduated in 1974 with a major in broadcasting and a minor in marketing. As a student, he dreamed of becoming a movie director. During senior year, students were encouraged to do internships in their chosen fields. Unable to find a job with local broadcast television, he decided to try cable television, which was then a very new technology. Not many people knew what it was, but Gordon had studied it in college and understood its many possibilities. Legally, cable television companies had to provide access channels that allowed local people to produce their own television shows. Gordon hoped he could get a job in local access television production, and perhaps work his way up, finally getting to Hollywood.

When he applied, though, he learned that the company, Warner Cable Television, part of Warner Communications Corporation, only had openings in sales. "Can you sell?" the manager asked him. Selling was something the twenty-one-year-old had been doing all his life. He took the job and was assigned to knock on doors in apartment buildings in gritty Chelsea, under the Mystic River bridge.

The blue-collar town next to Boston had very poor television reception because of the bridge itself and because of interference from Boston's tall buildings. Some college kids might have been deterred from working at night in Chelsea, a tough town, but to Gordon it was just part of what he'd grown up with. In the first week, after working about twenty hours, he had signed up fifty-three subscribers and made $401, an amount that astounded him for the seemingly short workweek he'd put in. Company executives were also astounded. Out of the 143 Warner cable systems then extant across America, no one had ever signed up that many subscribers in one week. Executives from corporate headquarters in New York City came to study the young salesman's pitch. Three months later, he was helping field train other salespeople.

Gordon had learned another life lesson: Rolling out a new technology can bring success and profits. He liked working for Warner, and he liked the concept of cable television, but he was already learning that he didn't want to work for a megacorporation, that he was a guy who definitely wanted to call his own shots. Like his father before him, he wanted to start his own business.

In May of 1975, while still working for Warner, Gordon was sitting in a two-block-long gas line on Brighton Avenue, in front of one of his dad's stores. Beginning in 1973, oil price shocks had begun playing havoc with the nation's economy. Caused by a variety of problems ranging from a Middle Eastern oil embargo to a revolution in Iran, the shocks resulted in long lines, tedious waits, and much higher prices. Few Americans understood the geopolitics involved, but they knew they were paying at the pump, whatever the reasons. From late 1973 to the spring of 1974, for example, average U.S. gasoline prices rose from about thirty-eight cents to about fifty-five cents—a seemingly incredible peak to American consumers.

Customers in that Brighton Avenue gas line were annoyed, but for Gordon, the annoyance created a eureka moment. He would start a company that marketed energy-saving devices. Newspapers were full of stories of families and of businesses that were severely stressed by the sudden jump in energy prices. He thought he saw an opportunity for a company that would focus on energy and make a contribution by helping individuals and companies reduce their fuel consumption.

In 1975, using his $3,000 life savings, the twenty-two-year-old Gordon hired a lawyer, signed incorporation papers, and founded Energy Management Inc. (EMI). The company would grow in sophistication and adapt with the changing times often over the next thirty years by paying close attention to a series of energy-policy initiatives that would be approved on the federal and state levels.

At first, EMI was purely a marketing company, almost a larger-scale version of Gordon's dad's corner stores. The company began by selling off-the-shelf gizmos and gadgets that improved energy efficiency and saved customers money. Many of these energy-saving items were tax-advantaged, allowing EMI to learn that tax breaks could be an important selling point in a deal. One of EMI's early sales was to the West End Club. With a large building to heat, the always-cash-strapped club had big-time fuel bills. Gordon and EMI sold the club a device that increased the efficiency of the building's furnace by helping the fuel burn more completely. The club got more heat for less money, which pleased them. An added consequence of the improved burning was that fewer pollutants were emitted. For everyone, it was a win-win situation. This early phase taught Gordon more business basics and also provided an entrée into many of the region's commercial and industrial facilities.

By visiting those facilities, EMI engineers realized that the company could provide a much more substantial service. There were obviously profits to be made chasing BTUs in New England's factories on a larger scale. The region's factories were old-fashioned and underengineered.

Lots of energy was wasted during manufacturing, particularly heat energy. In marketing their off-the-shelf products, EMI sales staff had spent long hours with factory engineers, who had described their plants' unique problems. EMI staff realized that they could custom design energy-saving systems that could improve a plant's energy conservation much more than would simply selling the plant predesigned off-the-shelf products.

Before the oil shock, few Americans thought much about wasted energy. Schoolchildren learned that the nation's resources were unlimited. There would always be more coal, more oil, more natural gas, there for the finding. Mid-twentieth-century concerns about fossil fuels focused more on the problem of air pollution, which was already a well-defined and fairly well-understood problem. But the 1970s price shocks placed energy on the agenda front and center as both a financial and national security issue. New England particularly suffered, because the region had no indigenous fossil fuels. Among the region's struggling factories was the New Hampshire–based Manchester Knitted Fashions, which manufactured polo shirts for Ralph Lauren and employed as many as 300 people. In textile manufacturing, energy is a primary expense. Each step of the process, from dyeing to drying, consumes huge amounts of energy in the form of heat. Frustratingly, much of the heat produced by burning fossil fuels escaped up a stack as waste heat.

Company owner Herman Werner was at his wit's end trying to figure out how to cope with the era's energy price spikes. A privately owned company, Manchester Knitted Fashions had no large resources to call upon for cash. Werner was having difficulty weathering the storm. When Gordon knocked on his door and started talking, Werner listened. Gordon told Werner that EMI engineers could find ways to take the wasted heat and channel it back into the manufacturing process, rather than simply allowing it to be sent up a chimney. That was music to Werner's ears, but he wasn't totally convinced. He invited EMI to study his factory's energy use, to interview employees, and to design some devices, but he declined to commit himself until he saw solid evidence that the promised savings would actually appear.

Gordon accepted the challenge. After weeks of research, EMI presented Werner with some very impressive numbers. Manchester Knitted Fashions had been using water from the Merrimack River, at 55 degrees, during the dyeing process. That water needed to be heated to temperatures sometimes as high as 220 degrees. The problem was, a lot of that heat was wasted in the process of boiling the water, just as heat is wasted in the process of boiling water on a stove. EMI showed Werner how to capture that wasted heat for reuse. The rechanneled heat raised the river-water temperature from 55 degrees to 110 degrees *before* the water entered the boiler, enormously lessening the amount of fossil fuel required—and the amount of money that would have to be spent.

Drying the dyed textiles also consumed huge amounts of energy, just as does a household dryer. Drying-room temperatures needed to be very high, and EMI found that about 240 degrees of heat was escaping up the textile-dryer stacks. Many people had thought about the problem of reusing that wasted heat but hadn't been able to overcome a major obstacle: lint. Just as a home dryer expels lint along with its excess heat, so does a commercial system. And just as excess lint can cause a house fire, it had caused factory fires.

EMI designed a system of heat-absorbent tubes that would catch the heat escaping through the stacks, but not the lint. The solution turned out to be simple: nonstick Teflon. Covering the heat-absorption tubes with that material allowed the heat to be caught, while the lint continued to float up the stack. The solution proved so successful that EMI patented it.

Gordon also showed Werner how, through forming a tax-advantaged limited partnership, the factory could save 35 percent in fuel costs, pay for the system with those saved costs, and even generate a small income stream with leftover money. In other words, Werner would essentially be paid to become energy efficient. Werner signed on.

As EMI became increasingly involved in designing these heat-recovery systems, the company hired mechanical engineers, electrical engineers, finance executives, and others. At its peak during this stage, EMI employed about forty-five people. With rave references from Manchester Knitted Fashions and other companies, EMI's future looked secure.

Then, once again, events in the Middle East changed the playing field. In the early 1980s, the Middle Eastern oil cartel fell apart. At the same time, American oil fields increased their production. Oil prices plummeted from forty dollars to about twelve dollars a barrel, reaching at least once an eight-dollar-a-barrel low. Once again, the motivation for energy-conservation systems vanished.

Gordon rolled with the punches. During the 1970s, Congress passed a package of forward-looking legislation intended to encourage a change in the nation's basic energy framework. Some of the legislation involved an attempt to begin to open up the long-stagnant system of electricity generation. In the late nineteenth and early twentieth centuries—the age of monopolies, big trusts, and Teddy Roosevelt's big stick—investors in electricity generation had claimed that the seemingly complicated business of providing the nation with electric light was intrinsically a "natural monopoly." They insisted that all functions of the electricity business—generation, transmission, and distribution—had to be handled by one vertically organized company. Competition, they claimed, would harm electricity reliability.

For decades, the idea that a small group of people ought to control the business of electricity held sway. State and federal governments exerted some control from time to time to greater or lesser degrees, but most people just paid their electric bill and didn't give the monopoly much thought. As long as their bills weren't too high and the lights worked, the public seemed to have decided, the men at the helm were doing a fine job. This created the acceptance of a Mayan-like "electrical priesthood"—a group of anointed geniuses, the fortunate select, who would manage the nation's electricity supply.

The electrical priesthood paradigm allowed the technology to become old-fashioned, risk averse, and, quite simply, collusive and stagnant. New ideas were not welcome. Innovation was ridiculed. Efficiency, of course, was simply irrelevant. Why worry about *that*, when all plant managers had to do was shovel in more coal or burn more oil, and the public would end up footing the bill. Under those noncompetitive circumstances, what had always been good enough would forever remain just fine. Even the "innovation" of nuclear power developed in the middle of the twentieth century depended essentially on the primitive nineteenth-century technology—developed well before science had even confirmed the existence of the electron—used by Thomas Edison and friends: steam spinning a steam turbine coupled to a generator.

It was as though the electricity industry was using an abacus when it could have been using a computer. Then the 1970s oil-gaming shocked policymakers into taking a second look at the problem of oil as a fuel source, which in turn got them wondering if the system couldn't be modernized even more. Could they ditch the abacus? But what technologies would be better? Maybe, they thought, relying on an electrical priesthood was inefficient. Maybe some genuinely new ideas, rooted not in the concept of boiling water but in some more modern applications of physics could bring about much-needed change. Maybe the promise inherent in $E = mc^2$ could finally be achieved.

Those policymakers decided not only that the system could weather the shock of innovation, but that the security of the nation, with less and less indigenous oil, demanded change. The 1970s legislation, crowned by a far-reaching energy-policy act passed under the Carter administration in 1978, began tentatively experimenting with change by gradually opening the electrical transmission and distribution lines to small power generators. Valuing small power production may not seem revolutionary, but it was in fact a 180-degree turn from the century-long trend of state control, and of bigger is always better, no matter what. There was, finally, a crack in the door.

The new legislation also encouraged innovation in energy production, particularly if that production used renewable resources. By that time, Gordon and EMI had become adept at finding and taking advantage of opportunities opened up by legislative initiative. EMI had become a get-there-first company, intentionally small, lithe, and adaptable. Gordon preferred that approach, rather than joining the herd yearning to grow to megacorporate size. What made him unusual, though, was that unlike many similar companies entering the legislatively opened niche, he took a long-term approach. Competition was important, yet the stability of working for years at his father's stores was also part of his character. He had an odd combination of patience and intensity.

When energy prices dropped and heat-recovery opportunities shriveled, EMI looked to the new policies. The 1978 legislation was so unusual that many young businessmen declined the challenge and its concomitant risk. Not Gordon. Other companies, including many of that era's wind-technology-development companies, took advantage of the tax breaks and tax shelters made available but never produced anything of public value. In a few cases, those entrepreneurs, in a perverse application of a well-intended law, ended up scamming the system by taking investment money and tax breaks and doing nothing at all.

Gordon, however, was precisely the kind of young entrepreneur the experimenting policymakers had hoped to nurture. Seeing his opening, at the age of thirty-two, young, crazy, and naturally ebullient, Gordon led EMI, now reduced to about a dozen employees, headfirst into the brave new world of independent power production. Gordon saw that there was money to be made in changing the stagnant system and he wanted to be part of that innovation. Almost simultaneously, EMI took on two goals, a short-term goal of building a wood-chip-fired plant and a long-term goal of building a gas-fired power plant.

First, EMI decided to build a 15-megawatt wood-chip plant in Alexandria, New Hampshire, a town of roughly 1,000 people located near the state's lake district. New Hampshire's public utility had put out a call for a small power project to be fueled by wood chips. The state's huge logging industry had a problem with treetops abandoned on the forest floor after the trunk had been taken for lumber. The treetops played havoc with ecosystems by encouraging the wrong kinds of plants to grow, by discouraging the right plants from growing, and by providing easily set-ablaze tinder during dry summers. Forest managers suggested chipping the slash and using it to generate electricity.

Tempted by the utility's promise of a long-term power contract, Energy Management decided to take a bite of the apple. Their earlier project-financing experience had given them an in with Boston banks. Armed with a long-term power contract from something as dependable as a public utility, EMI hoped to swing the financing deal despite its complete lack of experience in the power-plant-construction field. It was a bold play, definitely the biggest project EMI had undertaken to date. But with nowhere to go in the heat-recovery business, Gordon and company took the plunge.

The stumbling block for EMI turned out not to be bank financing. Boston's State Street Bank—pleased with the Manchester Knitting deal, as was Werner himself—was happy to do business again with EMI. The dependable revenue from the utility contract would provide an income stream with which to pay back the $25 million loan, so that the bank worked closely with the young developers, guiding them through the thickets of a major financing deal. Still, developing the plant was a risky endeavor. At one point, the thirty-two-year-old Gordon had two lines of equity on his Back Bay condominium. But the gamble eventually paid off.

EMI gained immeasurable experience, particularly in dealing with companies providing the fuel source. After all, if the wood-chip-delivery companies failed to provide their resource, the plant couldn't provide electricity, thus failing to fulfill its contract with the public-service utility. During that era, Gordon learned a great deal about the difficulty of having to rely on a long train of providers in order to generate electricity.

Unexpectedly, the biggest problem turned out to be community relations. Nearby wealthy summer folk in Hill, New Hampshire, opposed the plant because of its impact on traffic (there would be many wood-chip-delivery trucks coming and going over local country roads) and air quality.

Officials held a public meeting to discuss the proposal, the first such meeting EMI ever attended. All its earlier projects had involved working with industries that had already-existing plants. Gordon and company were about to learn that creating something completely new carried its own unique set of problems. EMI executives arrived at the Grange Hall to find people from both towns standing outside the hall, exchanging rather heated words.

The atmosphere was tense as Gordon stepped to the podium. He tried to explain the benefits to local people. "We're going to use 180,000 tons of whole-tree chips annually," he said. A fellow with a farmer's cap on his head, wearing a work shirt with suspenders holding up his jeans, was leaning against the back wall, looking skeptical.

"There will be about twenty-five trucks a day bringing in the whole tree chips," Gordon told them. "The whole tree chips are going to bring in jobs," Gordon continued.

Finally the fellow in the back spoke up. "Tell me something, son," he said. "Where are you going to get all this chicken shit?"

Gordon looked at the farmer. He wasn't sure he'd heard correctly. "Excuse me, sir," he said. "What was your question?"

"Where are you going to get all this chicken shit?" the farmer repeated.

Gordon tried to decide if the man was somehow making fun of the project. He seemed earnest enough, but his question was pretty off the wall. "I don't understand what you mean, 'chicken shit' . . ." Gordon said. The whole room turned to look at the farmer.

"Where are you going to get 180,000 tons of poultry chips?" the farmer asked slowly.

The whole place burst out laughing.

"No, sir, not *'poultry chips.' 'Whole tree chips,'*" Gordon answered.

With the tension cut, the meeting moved on. Ultimately, the town gave EMI a permit to build the plant. With the permits in place, EMI got a $25 million loan. The plant went on line in 1986. EMI sold its interest in 1993.

While EMI was developing that plant, executives were also planning for the future. By the early 1980s, even as Ronald Reagan was scornfully tearing Jimmy Carter's symbolic solar panels off the White House roof, New England grid managers worried about the continued availability of the region's fuel sources. Coal, oil, and nuclear power dominated with almost no natural gas. Elsewhere, natural gas was the fuel of choice, because it both seemed plentiful and created less pollution. When planners put out a call for natural-gas plants, EMI execs raised their hands. There was, however, a problem: getting their hands on the natural gas. With none in New England, it all had to be imported via pipelines. Those contracts were managed by old-style energy companies who weren't ready to welcome an upstart like Gordon.

Gordon thought he'd found a path around that obstacle when he started reading about plentiful gas supplies 4,000 miles away in Alberta, Canada. Then began two years of trips to Calgary. Boston born and bred, when he traveled to Alberta, he felt like a pioneer panning for gold. Maybe he would strike it rich. Maybe he wouldn't. Doesn't hurt to try, he thought.

When Gordon first arrived in the Canadian Plains rodeo capital as a thirty-two-year-old kid, the Canadians almost laughed him out of the province. Why should we talk to you? they asked. A company the size of EMI seemed like a joke. The gas executives spent an awful lot of time looking out their windows enjoying the beautiful Canadian Rocky

Mountains while he pitched his company. It was the tetherball game all over again. He just kept going back.

Eventually he succeeded. Armed with those long-term contracts, EMI built three natural-gas-fired generating plants and became one of several "anchor" companies that helped bring a new gas pipeline, the Iroquois, to New England.

Eventually, the playing field changed once again as the utilities stopped offering long-term power contracts because they had enough power for the time being. On the other hand, new legislation and orders from the Federal Energy Regulatory Commission, FERC, decreed that electricity would enter the competitive marketplace just as had telephones, airlines, and trucking. The age of the electrical priesthood's natural monopoly was over.

The industry would instead, slowly but surely, open up to the free market. In a business already famous for risk, building the new plants, called "merchant power plants," raised the risk exponentially. The bottom line was that the huge investment required to build a plant was based on a bet on the price of fuel and electricity several years into the future.

The opportunity for big money coupled with the added challenge meant that Gordon couldn't wait to get involved. Energy Management began developing three more natural-gas-fired power plants, one in Rhode Island, one in Massachusetts, and one in Maine. No longer held back by regulatory policies requiring the independently built plants to be small, EMI's new plants were much more ambitious.

In Dighton, Massachusetts, Gordon built a 170-megawatt gas-fired plant that, without the security of a long-term power-purchase agreement, provided a genuine financing challenge. Many experts in those early days of merchant power believed that financing institutions would require a fifty-fifty debt-equity arrangement. Gordon was convinced he could create a deal much more to his liking. Once again he succeeded.

A power plant is subject to highly variable income streams, based on the extremity of the weather. During a particularly cold or hot period, a plant can expect to make a healthy profit. Conversely, if the temperature is perfect and the days are not short, less electricity is used. A plant doesn't have as strong an income stream. A merchant power plant without a long-term contract had the added burden of being able to show it could pay its debts, no matter the weather or the season, no matter what the demand for electricity.

Gordon eventually convinced both his natural-gas supplier and his financiers, Canada's Toronto-Dominion Bank and TIAA-CREF, to buy into an innovative financing deal. Normally, a power plant would pay its monthly fuel suppliers before paying its bank debt, but Gordon found a Michigan supplier willing, for a fee, to allow Gordon to pay its bank debt first. Then, if the plant had money left over that month, the gas supplier would be paid. If not, Gordon could postpone payment until later. Industry watchers doubted he could pull it off, but Gordon managed to talk all parties into agreement, and he managed to get nearly 100 percent financing.

In the summer of 1999, Energy Management placed five of its power plants, two small plants and the three new merchant plants, up for sale. The others had been sold previously. Eventually, all EMI's plants sold at their peak value, before it became widely apparent that the rising price of natural gas in New England would become a serious problem for merchant plants. To some, it looked like Gordon had a crystal ball. He and other EMI partners made a fortune. Gordon now owned a $3 million Beacon Hill home overlooking the old West End where his grandfather had nearly died in the slaughterhouse accident. He owned two $80,000 garage parking spaces, other accoutrements of entrepreneurial success, and a variety of small investments in real estate and start-up companies. He had begun salting venture capital into projects like Seahorse Power, an award-winning solar-powered trash compactor.

For someone who hadn't yet reached the age of fifty, and whose grandfather had barely been able to keep his family fed, Gordon was pretty well situated. "We don't need big conference rooms or the trappings of power anymore," he was quoted telling a *Boston Globe* reporter around that time. "It's what you *know* that counts. The most innovative deals today are structured on napkins, over café mocha." He was feeling pretty confident about himself.

He did, however, need something to do. Touting new technology had carried him this far, he figured. Why not look around for something totally different, totally new, and totally challenging. That's when he picked up the *Boston Globe* story about Brian Braginton-Smith's idea for energy produced by the winds of Nantucket Sound.

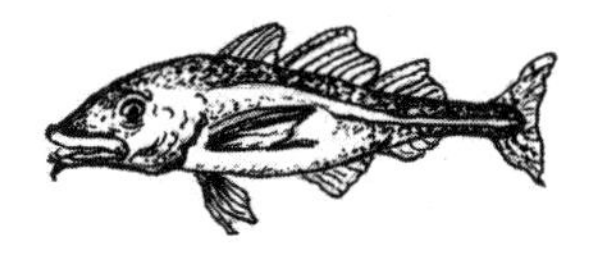

CHAPTER FOUR

Birds, Fish, and Whales

You furnish the pictures and I'll furnish the war.

William Randolph Hearst

Birds could be the Achilles' heel of Cape Wind, fixer Mark Forest began thinking hopefully. By December 2001, the project's wealthy opponents had stiffened their resolve. Word came down from above: Why have you guys let this project get this far? Forest began plotting how he could pull the region's heartstrings.

When powerful people start to dig in their heels and insist on getting their way, it's not always clear who is wielding the power, or who is writing the checks. In the case of Cape Wind, this was particularly true. Cape Cod's wealthy were more than happy to enjoy the limelight when the story involved attendance at an art gala or generous donations given to the local hospital or college, but when matters involved controversy, they hired underlings to speak for them.

On Cape Cod, this fact of life is buttressed by a century-old understanding that you don't bite the hand that feeds you. In return for the income brought by the wealthy folk, local people keep their mouths shut. Newspapers rarely write anything of substance about the region's superelite. Obituaries of servants who worked on Oyster Harbors estates might mention that fact, or a brief news story might run, as when a fire destroyed the $5 million, five-bedroom, four-and-a-half-bath home of Thomas F. Ryan Jr., the onetime president of the American Stock Exchange. In general, though, the elite were off-limits.

If few Cape Codders knew about Oyster Harbors—the club's name was most often mentioned in a hushed tone, if at all—the power of the gated community was no secret to Mark Forest. The onetime town manager of Provincetown on Cape Cod's tip, Forest, married with grown children, had been around the local political scene for decades. It was Forest who took on the task of strategizing Gordon's defeat.

Although by no means a scholar, Forest had a strong background in ocean policy, dating from his early career as an aide to the district's former congressman, Gerry Studds. By the time Studds retired from Congress in 1996, he had become known as a forceful advocate for the protection of the oceans. A national marine sanctuary, Stellwagen Bank, located in Atlantic waters between Cape Cod and Cape Ann to the north of Boston, was named after him. By the time of Gordon's proposal, Forest saw himself as something of an expert.

Marine sanctuaries are relatively new designations. Largely in response to an increase in the size and destructiveness of marine oil spills—oil was being transported in ever-larger vessels resulting in ever-more-destructive disasters—Congress under Republican president Richard M. Nixon passed in 1972 a wide array of new protections. One bill provided special status to marine mammals. Another prohibited dumping of such hazardous materials as radioactive wastes in coastal waters. Yet another allowed for the creation of special "sanctuaries" to protect exceptionally important or vulnerable underwater resources.

Sanctuaries guard such resources as endangered kelp forests off the coast of California and coral reefs along the Florida shoreline. Studds's

Stellwagen Bank preserves the shallow area's rich fishery and its plentiful population of playful humpback whales, which provide some of the world's best whale watching. When the Pilgrims first arrived in the New World, they saw so many cod that they wrote home that they could almost "walk across their backs." The cod is now nearly gone and the waters are woefully depleted, but the shallow bank is still nursery to many important species.

Marine sanctuaries protect more than ocean biology. The first marine sanctuary, sixteen miles off North Carolina, preserves the site where the USS *Monitor,* the Civil War ironclad, went down. Republican President George W. Bush designated a new region, the Northwestern Hawaiian Islands Marine National Monument, including the island of Midway, as a protected area.

While Studds was still the district's congressman, Nantucket Sound was proposed as a federal marine sanctuary. It failed to receive that coveted status because it lacked outstanding resources. The Massachusetts Office of Coastal Zone Management wrote in 1981 that the central sound "does not adequately meet site selection criteria for consideration" and that "adequate resources exist in Nantucket Sound; however, the majority of those resources are more readily definable in state waters and not in the central area of the Sound." Horseshoe Shoal was no Stellwagen Bank.

Since that time, the ecology of Nantucket Sound had continued to deteriorate. The shorelines are covered with estate homes, which polluted many of the Sound's estuaries and wetlands. Yachts and other large vessels routinely pump their bilges, filled with waste water, oil, gasoline, and other residues, into the Sound, so that small oil slicks are common during the summer months. Holding tanks filled with human feces are often emptied into the water. Project opponents regularly called the Sound "pristine," defined by Webster's as "still pure, untouched, uncorrupted, unspoiled," but pundits doubted those same opponents would drink the water of Nantucket Sound, were it fresh rather than salt. More and more frequently, the area's beaches—even Craigville Beach, "the fourth best beach in the world"—were sometimes closed because of high bacteria counts.

Nevertheless, Forest decided to dust off the old application. Although Nantucket Sound was unlikely ever to win national marine sanctuary status, merely claiming it was "on the list for consideration" would provide a hook for many a news story and create an exaggerated image of the body of water's ecological uniqueness.

Confusing the issue was the fact that the state had declared almost all of its *own* coastal waters, which extend three miles out to sea, an "ocean sanctuary." The two designations, the state's "ocean sanctuary" and the federal government's "marine sanctuary," are very different designations, established for two quite different purposes. National marine sanctuaries are so designated by virtue of their uniqueness. Conversely, almost *all* Massachusetts coastal waters are designated ocean sanctuaries. Indeed, the only state coastal waters that do *not* carry that designation are in and near Boston Harbor. The state's ocean sanctuary designation is not a matter of uniqueness, but instead addresses the use and management of near-shore waters and allows the state to play a role in matters like the construction and operation of coastal sewage-treatment plants.

The question of Nantucket Sound's national-sanctuary status, long buried, suddenly and unexpectedly resurfaced just before Christmas 2001, when Richardson and Blazis presented it at the last minute to the Barnstable Town Council.

The duo once again brought forward the resolution they had proposed in early October, this time with an added final paragraph: "The Barnstable Town Council does memorialize the Congress of the United States to designate Horseshoe Shoals [sic] specifically and Nantucket Sound generally, as a National Marine Sanctuary . . . ."

Reporters chuckled. The word "memorialize," the kind of language that would emanate most likely from a congressman's office, was definitely not town council verbiage. When Richardson admitted that he had "had some conversation with the congressman's office and was given some encouragement" to present the proposal, pundits understood that he meant he was doing Mark Forest's bidding.

Forest's strategy backfired. Town councilors, even some who opposed Gordon's project, were offended by the last-minute maneuver. For three months, a few councilors had been trying to research the pros and cons of the innovative proposal and had, at the previous meeting, begun to develop a long list of questions they wished to have answered. Now here were Richardson and Blazis, again asking for a snap decision.

"We all acted, I thought, in good faith," Councilor Richard Elrick said, adding sarcastically that it was "obvious that the issues are no longer relevant to us since we've already made up our minds."

Next came a voice out of the blue.

"I cannot believe this board feels it's doing its duty," said Councilor Richard W. Clark, an exceptionally articulate politician from Barnstable's Old Yankee north shore. This is "wrong! . . . You are denying me the ability to do what I was elected to do. You're asking us to do something that isn't right. We are elected to go through this the right way, the long way."

This was no act made up for television cameras. Clark was truly angry. The manipulation made his blood boil, particularly as it had already become apparent that the town council had little jurisdiction in the federal waters. As an avid scuba diver, he had spent as much time under the surface of Nantucket Sound as many of the project opponents had spent on the water's surface. Horseshoe Shoal was vulnerable to many scouring currents and was certainly not a delicate ecosystem, Clark knew. His diving had also made him very aware of the damage caused by commercial fishermen, who dragged their heavy nets across the shoal scooping up everything on the bottom in their search for a few species of marketable fish. As an engineer, he knew very well that the structures anchoring the turbines were likely to be less harmful than those draggers.

Clark stood up indignantly and walked out of the room in protest. Two other councilors followed him. Their action threw the meeting into an uproar. Richardson and Blazis backed down. Forest's proposition was tabled.

Forest began relying ever more heavily on the question of wildlife. Although rarely brought up in meetings during 2001, suddenly in early 2002, Save the Birds would become a common mantra. It was by no means clear that many birds visited Horseshoe Shoal, five miles out to sea, or that they would be harmed if the turbines were built, but the hyperbole would continue unabated for years.

Imaginations ran wild. A Web site called "NotCapeWind" claimed that the Sierra Club had called wind turbines the "Cuisinarts of the air." The Web site's perpetrator, John Donelan, then a part-time employee of the *Cape Cod Times,* refused to correct the error. By July, the Sierra Club's Matt Laskey called from San Francisco to insist that the comment be removed from the Web site. Even then it took a while, but eventually Donelan did so. During the same period, *Cape Cod Times* editorials, now virulently opposing the project, were accompanied by lurid drawings of seagulls about to get their beaks chopped off by wind-turbine blades. Long and impassioned speeches about birds were made in Cape Cod public meetings, and the term "Cuisinarts" found itself included in many a public record. It had a permanent place in opposition rhetoric.

The hysteria culminated in the writing of Jaci Barton, a woman who, like Royden Richardson, had close associations with the Osterville crowd. Barton was the executive director of the Barnstable Land Trust, an organization that had set aside many important natural areas for the public, but that also had helped put in place lucrative property-tax breaks for the Mellon family and for others in the Oyster Harbors crowd. For example, Barton helped work out a deal for land owned by Paul and Bunny Mellon's daughter, Catherine Mellon Conover, ex-wife and still close associate of U.S. Republican Senator John Warner of Virginia. (Warner and Catherine had several children together. Warner, who subsequently temporarily married movie star Elizabeth Taylor, owed his financial and political fortunes to the Mellon family and visited Osterville frequently. He was soon to become a rabid opponent of offshore wind energy, and, seemingly, of wind energy in general.)

Conover, a life member of the Barnstable Land Trust, had provided through various private foundations a great deal of money for land conservation in the town. She had also, through the Barnstable Land Trust, received a conservation easement on twenty-seven acres of her Oyster Harbors property near the DuPont estates. The easement provided Conover, the former Mrs. Warner, with a considerable property-tax reduction while still allowing her to build several new homes on her seashore land.

Given the rabid opposition to the wind farm by the Mellons and the DuPonts—two families whose decades-old presence on Cape Cod is rarely publicly discussed—it wasn't surprising that it was Barton's name that appeared on an opinion column claiming that Cape Wind would be a "killing field" for "a half-million" birds that she claimed flew over Horseshoe Shoal each year. Barton portrayed Gordon's field of wind turbines as creating "a habitat dominated by structures [that] could wreak havoc on fish, marine animals, and the organisms they feed upon." She subtly evoked Rachel Carson by writing that the turbines were being proposed for "the sea around us," the title of one of the famous environmental scientist's books. Then Barton charged that the Army Corps was "fast-tracking" the project and that "wind turbines don't work where the wind is unreliable and intermittent." She went on to romanticize Nantucket Sound, calling it "fragile." She claimed the project didn't "have the support of the environmental and wildlife coalitions or the vast majority of Cape Cod residents and visitors."

Barton's claims were absolute nonsense (the Corps, for example, had developed a lengthy and complex permitting process that would require several years of research), but to Gordon, that didn't seem to matter. He was summering on Nantucket Island with his family when the column ran. The energy entrepreneur had a daily ritual of walking down to a local café, where he sat with a cup of coffee, read the morning paper, and organized his thoughts.

"My God, now I'm the Pol Pot of wildlife genocide," he thought, after reading that he was proposing a "killing field."

Gordon, decidedly unfamiliar with the world of wildlife advocacy, knew next to nothing about the various animal and conservation groups in play on Cape Cod, which ranged from animal rights groups like the Humane Society of the United States (not to be confused with the American Humane Society, a respectable pet-protection organization) and the Cape Cod–based International Fund for Animal Welfare (whose president, Fred O'Regan, sat on the board of the wind-farm opposition group), to more responsible science-based groups like the Massachusetts Audubon Society and the Conservation Law Foundation.

And if Gordon didn't understand the politics of the various nonprofit groups purporting to protect Cape Cod wildlife, he knew even less about the wildlife itself. Although he would eventually have to spend more than a million dollars studying how passerines use Nantucket Sound, before Cape Wind he had never heard the term. He knew little about the migratory habits of the globe's songbirds, and in fact had given birds very little thought during his early planning stages.

That, he would learn, had been a big mistake. Opposition-funded focus groups had pinpointed the bird issue as one that would resonate widely, not just with Cape Codders, but with the nation at large. In America, birds are big business. A 2001 federal Fish and Wildlife study found that 46 million Americans, a bit more than one-fifth of the population, consider themselves bird-watchers. Of those, 18 million spend money to take trips to see birds. Birds enjoy a privileged constituency. The study found that bird-watchers are older, better educated, and wealthier than the average American. Of families with a yearly income of more than $100,000, 27 percent call themselves bird-watchers. The study found that bird lovers spent nearly $32 billion on retail sales connected to their avocation in 2001, creating almost a million jobs in the field and generating $85 billion in overall economic output. Oddly, 44 percent of Montana residents say they are interested in birds, while only 9 percent of Hawaiians admit to the activity. Interestingly, Massachusetts, despite its reputation as a dedicated environmental state, ranks toward the bottom, with 22 percent of residents claiming interest.

Cape Cod, an international destination for seashore bird watchers, ranks much higher than that, in part because the population is older and wealthier, but also because birds are a prevalent part of the Cape's natural surroundings. Watching an osprey dive into the saltwater and emerge with a foot-long fish in its outstretched talons is a common sight on the region's beaches. The fish hawk's hunt never fails to mesmerize even the most jaded tourist.

Very reluctantly, Gordon came to understand that he, too, would have to do site-specific studies. Even as animal rights groups spread rumors that Cape Wind would not have to undergo environmental review of any sort, Gordon and staff were negotiating with oversight agencies regarding the depth and extent of the research that would have to be done.

At first, Gordon hired several industry researchers who did literature surveys, compilations of earlier research that might provide baseline data regarding types of species that frequented the Nantucket Sound area. But officials from the Army Corps, federal Fish and Wildlife, and state agencies demanded more in-depth research.

The primary problem was that, although project opponents milked the heartstrings by complaining about harm to environmental icons like whales and seals and birds, there existed next to no data about which species visited Horseshoe Shoal, or about how the Shoal was used by those species. Were there, in fact, any whales on Horseshoe Shoal, or even in Nantucket Sound in general? Bunny Mellon's replenished barrier beach had become a nesting area for a growing population of terns and piping plovers, which the Massachusetts Audubon Society had been carefully monitoring. But did those important shorebirds actually fly out to the shoal, or did they feed elsewhere? Bounding Nantucket Sound was the Monomoy National Wildlife Refuge, home to a large number of seals and many important bird species. It was also a rest area for thousands of migrating birds in spring and fall. Did they fly over Horseshoe Shoal? If so, how low did they fly? Were they likely to collide with spinning turbine blades? Did they fly at night, when they could be confused by the project's navigational lighting?

No one knew the answers to such questions. There was no scientifically founded evidence to indicate that whales, for example, regularly used Nantucket Sound, and it seemed doubtful that such large marine mammals would frequent the shoal, which becomes so shallow during low tide. Whale watching is common on Cape Cod, but the vessels do not tour Nantucket Sound. As for seals, they do enjoy Nantucket Sound waters, but would they be harmed if a project were built several miles distant from their regular haunts? A considerable proportion of information regarding bird behavior comes from bird-watchers, who routinely report sitings through various organizations. Horseshoe Shoal was not frequented by bird-watchers, so little data existed, although the general disinterest of bird-watchers in the shoal could, in itself, be taken as a clue that birds were not common in those waters.

Officials did not want Cape Wind to turn into another Altamont. Clearly more information was needed. State and federal agencies began negotiating with Gordon over how extensive, and expensive, those studies would be. Gordon wanted to pay for a year's worth of data collecting, but federal Fish and Wildlife preferred three years, basing their wishes on the understanding that animals sometimes change their behavior from year to year. How much data would be gathered would be an ongoing point of negotiation throughout the permitting process.

Meanwhile, the Massachusetts Audubon Society had decided to conduct its own independent avian census of Horseshoe Shoal. The venerable organization, founded in 1896 to address the decimation of America's birds for the millinery trade, was one of the world's first conservation organizations. Run during its earliest decades by the state's wealthy upper class, the society was outstandingly successful in its wildlife-protection endeavors and in habitat conservation. Its inception brought about the founding of the National Audubon Society, although the two organizations are today separate.

In Massachusetts, when birds are at issue, the Massachusetts Audubon Society is the accepted nonprofit authority. As regards Gordon's proposal, the society was in a tough position. Some of the same families that had founded the powerful protection group were now, more than a hundred years later, vehemently opposing the offshore project. Project opponents immediately began citing Audubon as an opposition group, although the agency had not made such a public statement. A case could certainly be made that turbines harm birds, but another, perhaps even more powerful case could be made that avoiding fossil-fuel consumption by building wind turbines could greatly improve bird habitat.

Faced with a difficult choice and hoping to remain credible, Audubon opted to delay definitive statements while looking into the issue scientifically. There was, however, a considerable amount of backroom pressure from its Cape Cod financial backers, who were trying to force the organization publicly to oppose the project. Using a grant provided by the Massachusetts Renewable Trust Fund, the society began several years of bird census studies.

CHAPTER FIVE

A Pride of Pastels

We can have democracy in this country, or we can have great wealth concentrated in the hands of a few, but we can't have both.

U.S. Supreme Court Justice Louis D. Brandeis

While the Massachusetts Audubon Society was busy trying to design avian studies of Horseshoe Shoal, Nantucket Sound yachtsmen were busy with a project of their own. And so it was that on a deliciously warm June day in 2002, under a gentle blue sky, a pride of pastels gathered at the Wianno Club's historic old building for an early-morning emergency meeting. Women with pink Lilly Pulitzer shorts and pearl earrings and men with cocktail-cherry-red complexions and little yellow whales on their green trousers gathered on the front porch of the Nantucket Sound property.

They had worried expressions. Fliers had gone out: Come one, come all. Club members: Please bring guests. All Wianno Avenue was abuzz with indignation.

This was the worst crisis to hit the summer crowd in decades, even more menacing than Prohibition. They entered their situation room, a ballroom converted for the occasion to a conference room, with rows of folding chairs set up to accommodate the anxious members. The master of ceremonies spoke briefly, then introduced Douglas Yearley, the new head of the Alliance to Protect Nantucket Sound, formally organized only a month earlier.

The Wianno Yacht Club Commodore introduced the opposition group's leader by presenting Yearley as a staunch conservationist who belonged to a long list of "conservation groups of the highest quality," such as the Nature Conservancy, one of the nation's premier land-conservation groups.

"I think you will find him fired up and ready for action . . . . I have total confidence our goal will be accomplished," the commodore said.

In fact, Yearley was anything but a conservationist. Named "Copper Man of the Year" in 1993 by the Copper Club for his leadership in the mining industry, Yearley had spent his career with Phelps Dodge Corporation and had only recently retired from his position as CEO. Founded in 1834, Phelps Dodge markets itself as having helped "tame" the Wild West, but many modern environmental organizations would characterize the company's mining practices rather differently. The company led the way in open-pit mining and in copper smelting, and by the end of the twentieth century Phelps Dodge was fighting off a spate of legal cases stemming from the leaching of cyanide and other dangerous compounds, from failing to report toxic releases, and from noncompliance with federal legal requirements. Phelps Dodge had "polluter awards" from several environmental organizations and had paid large fines for environmental violations. In Tyrone, New Mexico, at a very large Phelps Dodge copper mine, beginning in the year 2000, birds had died near a water impoundment site found to have acid runoff. As Yearley walked up to the podium to speak about his environmental concerns regarding Nantucket Sound, the problem of bird deaths in New Mexico continued. (In 2005, the company would plead guilty to a misdemeanor in the case and pay a fine.)

Yearley also sat on the board of directors of Marathon Oil Corporation, once part of John D. Rockefeller's Standard Oil Trust, and a company with its own share of environmental battles to fight. Marathon Oil, involved in a Securities and Exchange Commission (SEC) investigation into allegations of bribery in Equatorial Guinea, where it was building a liquefied natural gas facility, would soon have its hands full.

After Yearley retired from Phelps Dodge in 2000, he began spending more time in his Wianno Avenue home, not far from the exclusive Wianno Club (which once upon a time never took Jews or Catholics) and only a few minutes' walk from the home of *Cape Cod Times* publisher Peter Meyer. Yearley's home was a bit more upscale than Meyer's. Sitting on almost one-and-three-quarters acres of prime waterfront land, the house, built in 1875, and the land, appraised together at $6.7 million, had several fireplaces and a $9,000 tennis court. The house had more than 7,700 square feet of living space, yet when *Herald* columnist Cosmo Macero interviewed him, the mining mogul explained that he was a serious conservationist interested in "sustainable living." Macero did not ask Yearley how much electricity was required to power such a large home, but he did ask about Phelps Dodge's not-very-respectable record in New Mexico. "Have you ever been to New Mexico?" Yearley reportedly responded. "I consider Nantucket Sound in the same league as the Grand Canyon. Without insulting people in . . . New Mexico, we are not dealing with that kind of beauty there."

Yearley had only bought his shoreline home in 1997, making him a newcomer to the Osterville scene. Why he assumed leadership of the anti-wind-farm group never became clear, as he was certainly not a very useful front man. But Yearley did have a dog in this fight. He and his guests would on clear days see Jim Gordon's wind farm on the southern horizon, generating electricity to power all those huge houses. Wags suggested that, to a man sitting on the board of one of the world's major oil corporations, the wind farm might have seemed like a personal insult, as though Yearley's time had come and gone, and the generation of Jim Gordon was only getting started.

The introduction by the Wianno Yacht Club Commodore having been completed, Yearley, sixty-six, walked to the podium amid a spattering of polite applause. He began by introducing his two recent hires for the Alliance, Isaac Rosen and John Donelan. Rosen had a small Cape Cod public-relations company and had been involved in several small-town political campaigns over several years. Donelan worked for the *Cape Cod Times* and also had a side business designing Web sites. It was Donelan who had set up the Web site that claimed that the Sierra Club opposed wind turbines because they were "Cuisinarts."

"I will be heading the legal committee because that is the most sensitive of our activities as we go forward," Yearley told his audience. The grassroots committee would be headed by John O'Brien, he said, and the environmental committee by Barbara Birdsey, an animal advocate and a Nantucket Sound homeowner.

Next, Yearley told his audience that he was often asked why he was taking the job as "CEO" of the "grassroots" Alliance. "When I retired, I wanted to give back . . . . This is a beautiful resource that we have here," said the mining executive, adding that he wanted it "for my kids and my grandkids unspoiled." Then he began his PowerPoint presentation.

"We aren't calling it a wind park. We're calling it an industrial complex," he said, pointing out the power of a sound bite. "Twenty-eight square miles, only four-and-a-half or five miles from Osterville. Put that in perspective."

Yearley then introduced Guy Martin of Perkins Coie, "our newly retained Washington law firm." Perkins Coie had long represented Phelps Dodge and a number of fossil-fuel companies. Martin himself had been intimately involved in Alaskan affairs since moving to that state in 1967. When he came to Washington, he was instrumental in helping to permit the Alaskan oil pipeline.

"It's a monster project," Yearly went on, claiming there would be 680 flashing navigational lights, foghorns, and other safety devices. "We'll have a continuous light show, visible day and night." There would be noise coming from this, the "largest construction project" on the Cape since the Cape Cod Canal. (In fact, the project would have fifty-seven flashing red lights atop the turbines, required by the Federal Aviation Administration, that would, according to the *Boston Globe,* look like "a faint row of holiday lights." The foghorns would be audible to a distance of roughly one-half nautical mile, but few people on Cape Cod complain about the sound anyway. If you don't like foghorns, Cape Cod is probably not the best place for you to live. During the six-month construction period, there would be noise, but following that, no one on land would hear anything.)

"We're not against renewable energy . . . but we're not for it in this location," he said.

"Who is Cape Wind?" Yearley asked, as though he were at some high-level strategy meeting, about to reveal some closely held secrets. "We know it has Italian money in it."

Hearing this after the meeting, Gordon would be confused. What was Yearley implying? Did he mean the project would be paid for in *lira*? Did he mean Gordon was actually from Italy and only *pretending* to be from Newton, Massachusetts? Then he laughed. There was Protestant money in his project, Jewish money, Irish money—but *Italian* money? "I love Italians," he would say, "unfortunately, that's the one ethnic group that *isn't* part of our investment group."

About environmental hazards, said Yearley, "there's quite an array here and the environmental community is quite active . . . although in the early going, Cape Wind has one major environmental group with them, and we have to wrestle that away from them." Presumably he meant the Boston-based Conservation Law Foundation (CLF), famous for its aggressive protection of the ocean and of coastal areas. CLF had already come out as strongly supporting Gordon's project, with the proviso that the regulatory review process show that any negative environmental impacts would be negligible.

"This is not a joke. They're deadly serious. They are well connected politically. They have a focus campaign that was very effective during the winter," Yearley warned.

There was no focus campaign. Yearley might have been referring to a Cape Wind–financed telephone poll run by Opinion Dynamics out of Cambridge, Massachusetts, which claimed that a slight majority of Cape Codders favored the wind farm. There certainly was no big-time for-hire spin machine making its way through Cape Cod meeting halls, although Gordon and his staff did line up as many public-speaking engagements as possible. Eventually he would hire Larry Carpman, a public-relations consultant, at $6,000 a month. Carpman, however, would not act as a spokesman for the project, but as a media adviser to Cape Wind. The company itself would continue to speak directly with the press.

Next Yearley brought up Massachusetts's senior senator Edward Kennedy, first elected to the Senate in 1962, and not a man to be ignored by the state's politicians.

"I've been through enough of these to know that wealthy people who are worried about views will not be the reason for this to succeed," the mining mogul said. "We cannot afford to be complacent. Ted Kennedy is not going to fix this for us. In fact, he's somewhat in hiding at this point, for reasons that are political . . . and if we don't go forward, this will go forward. We have to defeat it."

Yearley and lobbyist Martin stood in front of the audience, chatted and fielded questions.

"Guy," the mining mogul said, "give us the benefit of your wisdom from Washington."

Martin had been an aide to Alaska's Democratic congressman Nick Begich from 1971 to 1973. Begich died in a plane crash while campaigning and was replaced by Republican Don Young, who had held the seat ever since. "This project has what in Washington, D.C., we call traction. There are people in Washington who wouldn't mind seeing this project succeed," the lobbyist told the audience. Even worse, Martin complained, "the Corps of Engineers is coming at it pretty hard."

Most of the several hundred people in the audience appeared to buy into Martin's alarm call, but a few asked some searching questions. "Has the Union of Concerned Scientists [UCS] taken a position on this?" asked an older man. Yearley said he didn't know, but his guess was that they hadn't.

In fact, UCS *had* entered the fray, specifically saying that Cape Wind would provide a partial solution to the climate-change crisis. At that time, UCS's Deborah Donovan was writing an article for the fall issue of the organization's magazine. "It's human nature to resist change and to fear the unknown," she wrote, "but it is our hope that the residents of Massachusetts and the state and federal decision makers responsible for approving these projects will give these [renewable energy] opportunities a fair and rigorous review."

Lobbyist Martin broke in. "We're obviously going to work the politics of this as well as the legal aspects," he said. "You have no choice but to move your delegation."

Yearley explained that Massachusetts's politicians would need a lot of cover. In speaking with Congressman Delahunt's office (those in the know heard "Forest"), Yearley had learned that the state's delegation had a problem. Because, under Kennedy's leadership, the state's elected officials in Washington had opposed drilling in Alaska's Arctic National Wildlife Refuge, "there is a sensitivity about being too public," he said, "so we have to provide ammunition for the delegation . . . in the form of grassroots, environmental and other . . . so that they can step up and say

'My constituents don't want this and therefore I'm against it,' rather than 'I don't want it because I don't like the way it looks out in my front yard.'" To provide that ammunition, Yearley said, the Alliance would pay for a massive public-relations campaign that would create the image of a "grassroots" environmental effort. He introduced the Boston communications expert he had hired to do the job. It was Ernie Corrigan.

After the meeting, checkbooks were drawn from holsters. High noon for Jim Gordon seemed to be just around the corner. Women milled about on the club's lawn, chattering about how horrible those awful lights would look. What's the point of having a seaside verandah, they wondered, if there are lights everywhere? What about property values? If their views were to be ruined, who would compensate them? Evening cocktails, they remarked, will never be the same.

Later that afternoon, Jay Cashman, one of the Western Hemisphere's largest marine contractors, was fueling his boat at an Osterville dock when an acquaintance walked over. "We just raised $4 million over at the club this morning to stop that wind-farm thing," Cashman's acquaintance said.

Cashman didn't bother to answer. He was intrigued. He himself had just started thinking seriously about developing his own offshore wind farms. Turbines could be a great way to make money, he thought, and to help free the country from its dependence on Mideast oil. Right now, we're just rolling wheelbarrows of money over there, Cashman thought to himself, and look what we're getting back for our dollar bills.

If Yearley did indeed raise $4 million that day, the public will probably never know who signed most of the checks. Although the number of nonprofit entities has blossomed over recent years, IRS oversight remains minimal, to put it kindly. Often the names of donors to this growing number of "charities" do not have to be provided to the federal agency, and in cases in which the names must be reported, those names do *not* have to be released to the general public.

Additionally, money very often flows from one tax-exempt foundation to another, and then perhaps even on to a third. It can become a way to launder money. It's next to impossible for the public to discover the wizard behind the curtain—the ultimate source of the money funding the debate. And so it was with the Alliance to Protect Nantucket Sound. Despite repeated questions from journalists, the secretive organization refused to reveal the names of its big-money contributors, claiming that the funders of Yearley's "grassroots" movement deserved their privacy.

Records searches revealed probably only the tip of the iceberg—but it turned out to be a very interesting tip. Initially, project opponents funneled money through Three Bays Preservation, a nonprofit entity that did some legitimate environmental research and conservation in Osterville waters, but also dredged boat channels around Oyster Harbors.

In 2002, the family foundation of Richard J. Egan—Massachusetts Republican Party boss and financier, and close confidante of Mitt Romney—and sons donated $16,000 to Three Bays, specified for use by the Alliance to Protect Nantucket Sound. Later, the Egan Family Foundation listed a $2,500 donation to Three Bays, specified for use by the Alliance, a $90,000 donation directly to the Alliance, and a $100,000 donation to the Beacon Hill Institute. The Institute was run by David Tuerck, described by New Hampshire Governor Jeanne Shaheen as a "right-wing economist for hire." Tuerck's "institute" subsequently released an analysis of doubtful quality claiming that Gordon's project would cause significant economic losses to Cape Cod because of decreased property values and lost tourism. Tuerck's report was slammed by press outside of Cape Cod. Wrote *Boston Herald* columnist Tom Keane: "the Beacon Hill Institute is losing its credibility."

For the fiscal year ending in September 2004, the Egan family foundation listed a $300,000 donation to the Alliance, the highest the foundation gave that year and more than the foundation gave to Boston's Children's Hospital. In comparison, the foundation gave $1,000 to Rosie's Place, a shelter for homeless women in Boston. Richard J. Egan was cofounder of EMC Corp., a technology company said to rank among Massachusetts's most valuable. *Forbes* magazine estimated his personal fortune at about $1 billion. Through a series of trusts, his family owned a list of properties on or near Nantucket Sound. They told Gordon that they had no intention of having to look at his turbines when they wanted to visit their homes.

Another major Alliance contributor was Paul Fireman, who would be ranked the world's 698th richest person by *Forbes* in 2006 after selling his company, Reebok, to Germany's Adidas. Fireman invested in Reebok in 1979, when the company was still British. He eventually made the brand popular with U.S. athletic teams, hence with American sports fans. Fireman bought up a considerable amount of acreage in the town of Mashpee, located to the west of Osterville, and built his own golf and tennis club. Willowbend, not particularly welcomed into the Mashpee community, was a notorious bad neighbor: it erected a high fence that created an eyesore; waters along an abutting estuary became polluted after its golf course was built; and it insisted on running its loud machinery well into the wee hours of the night, occasioning angry phone calls from people trying to sleep and visits from local police. Although he was selling homes in Willowbend, Fireman built a separate house for himself in Osterville on Nantucket Sound near Oyster Harbors. Valued at more than $6.3 million, the 10,000-square-foot house, a kilowatt sink, on 3.3 acres is anything but environmentally sensitive.

In 2003 the Paul and Phyllis Fireman foundation gave the Alliance $50,000. That figure increased in 2004 to $200,000. Fireman also agreed to hold a fund-raiser at Willowbend, with Yearley as the guest of honor.

Invitations to the event were sent to area residents, including to Chris Sherman, a Cape Wind attorney.

Sherman decided to attend.

When he signed in and had his name checked off, Alliance staff recognized him and realized the mistake they'd made. Heads turned. The whispering among several couples got louder and louder. Then Sherman saw an attractive older woman striding purposefully in his direction. "I f***ing hate you," said the woman, who turned out to be Jamie McCourt, now co-owner of the Los Angeles Dodgers with her husband, Frank.

Sherman, ever the gentleman and by nature rather mild-mannered, was shocked. He had never met the woman before. "You don't hate me," he said, soothing the irate woman. "You might hate my project, but you certainly don't hate me. You don't know me."

Alliance funds have also come from the Point Gammon Foundation, connected to the descendents of Malcolm G. Chace. Chace, a Providence businessman, owned Berkshire Fine Spinning Associates, which merged with another company to become Berkshire Hathaway, now headed by the legendary Warren Buffet and no longer a textile company. In the early part of the twentieth century, Malcolm G. Chace had bought up a good deal of Cape Cod's southern shoreline, which he marketed to the wealthy. During the 1960s, Chace had been involved in illegal dredging activity along the Cape's southern shoreline, which created a substantial number of high-end properties, now known as New Seabury. He was named as a coconspirator in the case, but was never indicted because he chose instead to testify against a state waterways director, who was convicted and sent to jail.

The Boston Foundation, an umbrella "community" foundation established in 1915, was a funnel through which a considerable amount of Alliance-bound money flowed. The amounts were small at first. R. Abel and Nancy L. Garraghan donated $1,000 in 2003. Garraghan was the head of Heritagenergy, a New York–based fuel-oil business.

The following year, the amounts contributed through the Boston Foundation were considerably larger. The Garraghans again contributed $1,000, as did Lucius T. Hill III. Hill, a contributor to many Massachusetts environmental causes, was the descendent of a Lucius T. Hill who was the "Commodore" of the Beverly Yacht Club, Beverly being a town of considerable means north of Boston. The Nancy and George Soule Family Fund contributed $7,500. The first Soule in America came over on the *Mayflower* as the servant of Edward Winslow, but didn't stay a servant for long. The Atlantic Fund contributed $30,000, but the derivation of that fund remains unclear, as the Boston Foundation does not have to provide to the public the names of the people behind it.

The Albert J. and Diane E. Kaneb family fund donated $130,000. Al Kaneb was once president and co-owner of Northeast Petroleum Industries, an oil company. He also owned Barnstable Broadcasting, a firm that owned a very large number of radio stations across the country. The Kanebs owned a Nantucket Sound property valued at more than $4 million, not far from the Wianno Club and from Marathon Oil's Doug Yearley. The Kaneb family was to donate even more, $168,250, through the Boston Foundation the following year, and Kaneb himself was to become an active fundraiser. Amounts from the other donors remained stable during 2005, although the Soule family did not donate again. Altogether, at least $400,000 was funneled into Alliance coffers via the Boston Foundation.

Other Alliance nonprofit foundation contributors included several connected to the DuPont family; the Edward C. Johnson Foundation, connected to Fidelity Investments; and the Manzi family foundation, derived from Lotus software. Sometimes very odd foundations contributed. In August 2002, the Boston-based Birmingham Foundation, which contributes almost exclusively to Catholic schools, gave a $10,000 check to the Alliance to Protect Nantucket Sound. The solution to that mystery may lie in the fact that the foundation's two trustees, Paul J. Birmingham and Lois I. Wrightson, live in Palm Beach and Osterville respectively. The Holloway Foundation, which normally gives to worthy children's charities like the Special Olympics, would contribute at least $7,500 to the cause. Gary and Julie Holloway, trustees of the foundation, owned a $6.2 million home on Squaw Island, the island on Nantucket Sound where Ted Kennedy lived for so long.

As Yearley continued to speak on that late June day, he described the Alliance's strategy over the coming years. He promised his contributors a multimillion-dollar campaign that would spend as much as $100,000 a month on his grassroots efforts and an equivalent amount on legal strategies designed to keep Gordon in court and off Horseshoe Shoal. The Alliance would ultimately spend well over $10 million trying to keep Gordon's project from being permitted, much of it spent on lobbyists and communications men like Corrigan.

Wags sometimes wondered if a prerequisite to buying Nantucket Sound property was the possession of a well-endowed family foundation. Nevertheless, although there were plenty of property owners on Oyster Harbors, in Osterville, and along the shoreline in general who were willing to tap their charity funds to fight Gordon's proposal, the Alliance would find itself hard-pressed to keep up with its mounting bills. For help, it turned to a New York City-based professional fund-raising outfit, Community Counseling Service (CCS).

The benignly named organization was anything but a counseling service in the generally understood sense. Rather, it was one of the most popular professional fund-raising companies in the business. Nonprofit organizations pay these professional fund-raisers to raise money for their organizations, and in return the organization charges either a percentage of whatever money it raises or a flat fee. Community Counseling Services claims to retain about three to eight cents of every dollar brought in, a relatively small percentage. Some fund-raising services keep as much as 15, 20, or even 25 percent. Interestingly, although CCS is the organization doing the fund-raising, the person writing the check might never hear their name. CCS does its best to stay in the background, so that the donor never knows of its role.

Community Counseling Service has raised funds for some of the best-known charities in the United States, including Habitat for Humanity, Rotary International, Lions Club International, the Nature Conservancy, and even the Conservation Law Foundation, a Boston-based environmental advocacy organization working hard to support Gordon's project. Some found the service's willingness to raise money for the other side to be a sad indication of the state of contemporary American democracy and of the business ethics of some who support the nation's burgeoning nonprofit industry.

Under CCS's guidance, the Alliance to Protect Nantucket Sound began to bring in many more millions in donations. The service coached Alliance employees in how to approach "million-dollar prospects," going so far as to write out the Alliance's fund-raising letters to such prospects and then to write out a script for Alliance fund-raisers when talking with their wealthy marks.

The solicitation letter to "John Q. Prospect," left on the Web for the public to read, bragged about all the lobbyists the Alliance paid for, including four different Washington lobbying firms; clearly stated that an important goal was to use the political process to have Nantucket Sound declared a national marine sanctuary; specified that it had hired letter and column writers to try to have anti-wind-farm propaganda placed on the editorial pages of the nation's newspapers; and focused on its influence with elected officials.

Included in the letter was a picture of Massachusetts Governor Mitt Romney standing on Craigville Beach with Susan Nickerson, an Alliance official. The photo clearly implied that a major donation to the Alliance could bring access to Massachusetts's highest elected official. Also included in the letter was a discussion about the Alliance's interest in "wresting control" of the permitting process from the Army Corps, which the Alliance had been frustratingly unable to influence. The letter closed with several pictures of yachts sailing Nantucket Sound.

Also available on the Web site was the extensive script CCS wrote out for the Alliance. "Be relaxed; begin with casual talk," CCS advised under its section headed "Breaking the Ice."

"Don't let them pre-empt you by offering a gift," the document continued. "Simply state that you appreciate their generosity, but want to take some time to share your vision and describe what the Alliance's plans are to ensure its ability to defeat Cape Wind and establish Nantucket Sound as a Marine Sanctuary." It advised Alliance fund-raisers to explain to prospects that the agency needed money to "employ more State and National strategists to carry our message to lawmakers." Another name for strategists, of course, is "lobbyists."

The document also instructed Alliance fund-raisers in how to respond to the prospect's decision on how much to donate. "Use your best judgment," the document said. "If it is a token gift of $5,000 you should delay acceptance."

When the documents were discovered, the press had a field day. Some thought the documents Machiavellian. The *Boston Globe* played the story on page one. The *Cape Cod Times,* however, did not report the Community Counseling Service story at all.

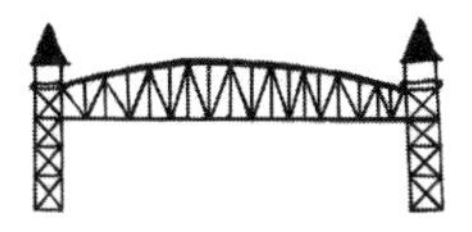

CHAPTER SIX

The Passion of Matt Patrick

It's a rough game, underneath the backslaps and the handshakes and the big noble speeches . . .

Allen Drury, *Advise and Consent*

The Community Counseling Service–generated letter to "John Q. Prospect" claimed that the Alliance had "gained support from *all* the local politicians" and had "developed a strong base of influential republican and democratic supporters in the State House and in Congress." Politically, it did look as though project opponents had everything sewn up.

Even Mitt Romney, close to Republican financier Richard J. Egan, had signed on to the cause. In early 2002, Romney announced he would stand for the fall gubernatorial election. Gordon sent Romney a campaign donation and attended a fund-raiser. When he approached Romney to talk about the project, the candidate openly admitted that he'd already made a campaign promise to oppose Gordon's project.

"I never go back on my promises," Romney told Gordon.

Gordon was astonished. As with Kennedy, Gordon had never had a chance to explain his proposal to Romney, yet the candidate had already committed himself to opposing a project that could bring a huge economic windfall to Massachusetts.

What's the deal? Gordon wondered. It may have shocked Gordon, but Romney meant what he'd said. Visiting *Cape Cod Times* editors, he repeated his stand. As far as ocean views go, the Cape need not worry, Romney told them. The ocean view from the Cape is a national treasure: "You don't interrupt that."

Lining up Cape Cod politicians had been one of the Alliance's first tasks. Between the efforts of congressional aide Mark Forest and the chamber of commerce's John O'Brien, many of Cape Cod's boards of selectmen had gone along with opponents. Additionally, the Cape's state senator, Rob O'Leary, and most of Cape Cod's state representatives had joined the bandwagon, opting to follow in the footsteps of either Republican Mitt Romney or Democrat Ted Kennedy.

But there was one holdout. Matthew Patrick, a Democratic state representative from Falmouth, refused to toe the line. Growing up, Patrick had idolized the slain Robert F. Kennedy. But when the Kennedy family opposed the wind-farm proposal, Patrick found a postcard of John F. Kennedy sailing his yacht in Nantucket Sound and sketched in an array of wind turbines on the saltwater horizon.

He displayed the touched-up postcard prominently in his office. At forty-nine, Patrick was an unusual politician. He genuinely had little ambition for power or for prestige. Friends often coaxed him to run for higher office, but he always said he was satisfied representing his small constituency in state government. "Now and then, an innocent man is sent to the legislature," once quipped Kin Hubbard, a Midwestern newspaper columnist and close friend of Will Rogers. He could easily have had Patrick in mind. The Falmouth representative had no interest in the glamour of Cape Cod's wealthy class, didn't care about partying with the glitterati, and was happiest fly-fishing. He owned a small runabout that could take him along the shoreline, but it wasn't a large enough boat to safely go any distance out into the Sound.

A lifelong Democrat, he had grown up in a small New Jersey town—Millstone, population 400—and was the oldest of ten children. His father was a teacher. His mother divided her time between caring for her children and serving as the town's mayor, a post she held off and on for

Massachusetts state representative Matthew Patrick nearly lost an election because he asked Cape Codders to stop to consider the merits before voting to oppose Jim Gordon's Cape Wind project.

twenty-five years, until the age of eighty. Patrick had lived a Huck Finn childhood, fishing the streams, camping out in the woods, swimming and floating down rivers to the sea, and wandering through farm fields. An all-American kid, he had a talent for watercolors and a championship record as a high school cornerback, gaining him a scholarship to Upsala College, without which he would not have been able to afford higher education.

He came to Cape Cod almost by accident. He and his wife, Louise, whom he met in Ghana while both were Peace Corps volunteers, had looked for a place to settle down that was affordable, and in the late 1970s, property on much of Cape Cod, other than along the central southern shoreline, was comparatively inexpensive. In Falmouth, Patrick began working as a carpenter and jack-of-all-trades, then started his own energy-conservation business. A Federal Housing Administration loan helped them buy their first small starter home. Two children were born.

Patrick became an environmental mover and shaker on the local scene, leading a fight to restore a local trout stream, protecting important waterways from overdevelopment and preventing an 800-unit condominium development from being built within fifty feet of an important local river. He led the fight to establish the Cape Cod Commission, a land-use regulatory authority. He was chairman of the Citizens for the Protection of Waquoit Bay, an organization that had successfully protected Washburn Island, an island that was slated for Oyster Harbors–like development but which instead became a state park open to all.

Patrick was also something of a local renewable-energy expert. He ran energy-loan programs for the public, helped create a heating-oil buyers' cooperative, helped produce an energy-management plan for Barnstable County, and was involved in efforts to erect modern wind turbines on a Cape Cod military base. He had spent a considerable amount of time helping a Cape Cod clean-air group pressure the Cape

Cod Canal power plant to install technology that would decrease its toxic emissions.

He was also tremendously stubborn. Once he made up his mind that something was "right," he felt compelled to speak out. His friends weren't surprised when, at the Barnstable Town Council, he was first to speak in favor of giving Cape Wind a fair shake, even before he knew which way the political winds were blowing. And later, when the town council called for national marine-sanctuary status for the Sound, Patrick spoke to the larger context, connecting the September 11 disaster to global-energy questions, and suggesting that Cape Wind could be one small step toward a solution.

The problem for project opponents was not just that Patrick supported studying the wind farm—a handful of other Cape Cod politicians also did so—but that he wouldn't shut up. When the *Cape Cod Times* published editorials with incorrect information, Patrick wrote to correct the record, charging that the paper was providing its readers with "questionable information."

When the chamber of commerce's John O'Brien misled people by saying that New England didn't need more electrical power, Patrick again stubbornly insisted on setting the record straight. Cultivating O'Brien as an enemy was somewhat dangerous. O'Brien had once physically spun Al Benson around and accosted him verbally for publicly contradicting misinformation O'Brien had given out.

Rumors began to spread that Patrick was not a "team player." Led by Congressman William Delahunt, the hazing began. At a St. Patrick's Day political roast, Delahunt ridiculed Patrick in a speech and called him "Mr. Wind." Another state representative, Barnstable's Demetrius Atsalis, a compliant wind-farm opponent, asked Patrick to wear a little beanie with a propeller on top. State Senators Robert O'Leary and Theresa Murray also chimed in with their little jibes.

Patrick was prepared. He and his son had written an essay: "Top Ten Reasons to Oppose the Cape Cod Wind Farm." Among those reasons:

- The wind turbines won't match the color of the new drapes in the boathouse we just redecorated.
- The turbulence from the wind wake caused by the stupid windmills will take the wind out of our sails.
- It will be damned difficult to get back to Hyannisport after I've had a few drinks at the bar over at Edgartown.
- My stock in the nuclear-power plant will go down the toilet.
- They will spoil the Figawi Race, and you know how much trouble those boys have getting home without anything in the way.
- The ambassador says it will spoil the view of the sound from his $13 million mansion he built in Cotuit after tearing down a house that only cost him $4 million.

Amusing as his list was, Patrick was making fun of some of his own constituents. Osterville and its yacht clubs were part of his district. Ridiculing your own constituents is usually considered a faux pas, but when those ridiculed are some of the world's wealthiest families, the joke could be seen as suicidal.

In particular, the joke about "the ambassador" was not appreciated by Osterville Republicans. The ambassador referred to Richard J. Egan, Mitt Romney's friend, a Bush Ranger, and the Bush administration's then representative in Ireland.

Alliance supporters decided to get even.

Once again strange things happened in the Barnstable Town Council. In May 2002, shortly after the political roast, town councilor Richard Barry made an announcement: "Mr. President, I'd like to, with your permission, introduce Larry Wheatley. He's a gentleman who's sitting out in the audience right now. He's going to be running against Matt Patrick. I'd like to afford him an opportunity to introduce himself."

Once again, councilor Richard Elrick spoke up, pointing out that introducing a future political candidate via the televised town council meeting was inappropriate, and that allowing him to make a televised campaign speech under the aegis of the town council was an outrage to the democratic process.

The council president interrupted Elrick. "It is unusual," the president conceded. He said he would take a vote. If he got eight votes, he would let Wheatley speak. He got seven votes.

He stared intently at several councilors, then polled them again, but still got only seven votes. This time, he raised his own hand, getting the required eight.

"Come aboard," the council president jovially ordered Wheatley. Wheatley rose from his folding chair in the audience and slipped his suit coat over his white shirt while walking to the podium. His thinning red hair was neatly combed over his forehead. His bright red tie with small white polka dots nicely set off his American flag lapel pin.

"Thank you. Thank you, councilors. Thank you, sir," he began. His voice fluttered, perhaps because he realized that a campaign speech in this forum was very improper. "Not to take the time to talk issues," he said, commencing to talk issues. "I know where folks in those areas stand on wind farms. I stand with them . . . ." Mission accomplished, Wheatley sat down.

Most of the councilors said nothing about the candidate's outrageous behavior, giving a campaign speech at the council that would be televised throughout the town of Barnstable. Elrick, though, was furious. "I think we will rue the day that we chose to interject partisan politics into this night's meeting or into any town council meeting," he said.

"It is unprecedented. There are opportunities for people to speak at two public comment sessions, both before and after this meeting. I think it was a mistake on the part of this body, and I think it was unfortunate

Larry Wheatley, the Republican candidate funded by Cape Wind opponents, who dragged successful political candidate Matt Patrick through the Massachusetts court system after Patrick won election to the Massachusetts House of Representatives.

and particularly unfair to Representative Patrick to not have an opportunity himself to be here tonight. We've never done this before, and I certainly hope we never do it again."

Republican candidate Larry Wheatley was a newcomer to Cape Cod who had once written a letter to a local paper claiming that Gordon's wind turbines would "endanger private pilots." A Coast Guard retiree and part owner of a couple of car washes, he said he was an attorney and he seemed to be selling real estate in Osterville. Other than serving on a local airport commission, he had no public political presence in the area.

Wheatley hammered away relentlessly at Patrick's support of Cape Wind. He held the press conference announcing his candidacy on a beach, dressed formally in suit, white shirt with cufflinks, tie, and ever-present American flag lapel pin. Several other similarly dressed men stood with him. Their spit-and-polished wing tips sank into the hot sand. His speech made clear that his main campaign plank was to join with other Cape politicians in opposing Jim Gordon.

Following his announcement, articles began to appear in the *Cape Cod Times* slamming Patrick and praising Wheatley. "Wheatley is taking on Falmouth Democrat Matthew Patrick, who could prove vulnerable as one of the few politicians east of the Hudson to support the proposed wind farm," jeered one reporter.

Massachusetts campaign-finance laws limit contributions by individuals to candidates for state representative to $500 per year. Nevertheless, Wheatley seemed to have plenty of money to work with. Campaign-finance records showed that the candidate was backed by Richard J. Egan, his family, and his business associates. Egan used a technique known as "political bundling." Bundlers like Egan control an extensive network of top-level contributors who can be called upon by the political boss to contribute the maximum legal amount to the boss's chosen candidate. Much of the money funding Wheatley's attack on Patrick came from the Osterville crowd. The Wheatley-Patrick battle became one of the most expensive state representative campaigns in Massachusetts history.

On election night, 2002, Patrick campaign supporters, gathered for the victory party, didn't quite know what to do. Patrick seemed to have won, but on the other hand . . . maybe he hadn't.

Throughout the long election night, as the crowd chowed down on appetizers and made liberal use of the cash bar, they celebrated, but not with abandon. As the count from the precincts came in, the lead seemed to be wavering, sometimes definitely in Patrick's favor, sometimes precariously close. Patrick's supporters kept hearing, via reporters, that the Wheatley camp was celebrating victory, too.

By evening's end, Patrick seemed to have won by twelve votes. Patrick nixed breaking out the champagne on ice. Let's wait until this is settled, he told his confused supporters.

"What happened?" his wife asked when her husband came home at the end of the long night. When Patrick told her about his tenuous victory, she asked what that meant. He answered that he didn't really know, but that he was sure of two facts.

"It means a recount," he said. "And it means we need a lawyer."

A bizarre legal battle began as Wheatley and his financiers relentlessly stalked Patrick through the Massachusetts court system over nearly a year, dragging him all the way to the Massachusetts Supreme Judicial Court. Cape Wind opponents had promised openly that they would use all available legal avenues to stop Jim Gordon, and voters eventually understood that they meant to extend that legal crusade to Matthew Patrick, who had had the temerity to stand apart from the rest of the political crowd.

Their pursuit was to backfire. When voters' sense of fair play was aroused, the wound would be mortal for Larry Wheatley. It would also mark, almost imperceptibly, the turning of the tide for Jim Gordon. When the Wheatley-Patrick controversy spilled over into the Boston arena, it made headlines in the Boston papers as well as elsewhere in the state. This brought the issue of the Cape Cod wind farm, hitherto considered by many to be a parochial issue, into the limelight. Off-Cape media became curious. People elsewhere in Massachusetts began to wonder what, exactly, was going on down there in vacationland.

The first step was a recount, and for that Patrick went to William A. McDermott. Billy, as everybody called him, had been hanging around the world of election law for a good many years. In Massachusetts, only a few lawyers knew election case law inside and out. William Galvin was one, but he was secretary of state. The powerful Massachusetts Speaker of the House, Democrat Thomas Finneran, a longtime Boston politician, directed Patrick to McDermott, who kept an office in a two-century-old house in the West Roxbury section of Boston. In the small, tight Boston political crowd, Billy knew anybody and everybody. Patrick needed someone like that. Other than when buying a house, he'd had few dealings with lawyers. Just the need for a lawyer made him uneasy, but McDermott had a well-honed skill for soothing anxious clients.

Billy was about as Boston as they come. He grew up in Savin Hill, an area settled in 1630, as old as the city of Boston itself. He graduated from Boston Latin School in 1963, Boston College in 1968, and Suffolk University Law School in 1973. After finishing law school, while working in Boston mayor Kevin White's administration, McDermott became a Boston election commissioner and decided that it would become one of his specialties. When McDermott began as commissioner, police officers took the city's census by knocking on residents' doors each year, a rather inefficient use of police time. McDermott helped design a yearly mailed census that saved the city money and was much less intimidating to residents.

"It's just a matter of odds," McDermott told Patrick. "If you're already ahead, chances are, you're gonna pick up a few more votes."

From the moment a recount petition is filed, the event becomes highly charged. Both sides maneuver to get their own favorites on the recount team. Each side gears up for war.

The recount began on November 23, 2002, in the town of Mashpee. Voters might imagine that a recount is an orderly and organized affair, that the democratic process at work is measured and judicious, and the vote is either clearly for Candidate A or for Candidate B. Nothing could be further from the truth. The recount room, stuffed with well over 100 people, was filled with tables at which sat Patrick and Wheatley supporters. The two candidates walked from table to table, trying to keep their supporters focused on the job, in the midst of the mayhem. At the registrars' table, dubious ballots underwent a kind of mini court session to determine which candidate had been voted for. If agreement could not be reached, registrars put the questionable ballot into a sealed envelope for a future, more thorough examination.

By day's end, Patrick had picked up three votes, but there were still three other towns in his district. All those votes had to be counted. Anything could happen. But after the final recount session in Barnstable Town Hall, Patrick had increased his lead over Wheatley to seventeen. "The voters have spoken," Patrick told a newspaper reporter. As far as he was concerned, the contest was over.

Hearing the final recount tally, Wheatley retired behind closed doors with many of the same Barnstable town councilors who had opposed the wind farm and who had put the Republican up for election against Patrick. When he emerged, Wheatley had a public announcement. "We are not ready to concede that the election process is over," he said. Once again, Patrick nixed breaking out the champagne.

Patrick waited for Wheatley to file his promised challenge, but it was not forthcoming. More than three weeks after the recount, on December 18, Massachusetts Secretary of State William Galvin certified Patrick as the winner.

Unexpectedly, on that same day Wheatley filed for a new election, mystifying McDermott. The law didn't specify a definite time limit for filing, but the way McDermott read the law, once the election was certified, the courts lost jurisdiction. Article 10 of the state's constitution clearly said that the House decided whom to seat at that point. Maybe Wheatley, being from Virginia, just didn't know the election law, thought McDermott. On the other hand, the man did have Massachusetts attorneys. In all of McDermott's experience, candidates contesting an election filed eight or ten or maybe eleven days after the recount. He'd never seen anyone wait this long.

Wheatley's filing contested a handful of ballots in the town of Bourne, as well as a very short closing of polls in Osterville, where Wheatley had won by a better than two-to-one margin. In Bourne, where Patrick had won, at one point the wrong box of ballots had been delivered to the polls, so that a small number of voters received ballots without the names of Patrick and Wheatley. In Osterville, when incorrect ballots had also been sent out, a policeman had told waiting voters that the polls would close temporarily until the correct ballots arrived. Wheatley claimed this delay lasted forty-five minutes, causing "many" people to leave without voting. Had they stayed, he claimed, he would have won. But the policeman and the election supervisor estimated the wait at only fifteen to thirty minutes. The policeman filed an affidavit saying that no one told him they would be unable to vote because of the wait. Nor did anyone appear saying he had not voted because of the delay.

A legal Ping-Pong game began, as motions and countermotions were filed. Then on December 30, two days before Patrick was to be sworn in for his new term, a Barnstable superior court judge decided a new election should be held.

Matt Patrick may not have been familiar with Oyster Harbors, but House Speaker Finneran certainly was. As a boy in the 1950s, Finneran had helped with his dad's carpet-cleaning business, which had several Oyster Harbors folks as customers. He'd been over that island drawbridge many times in his early life. As a politician, he had come to know some of the island's powerful people quite well.

Early on, Gordon had briefed Finneran on his proposal, hoping to earn the symbolic blessing of the influential speaker. Finneran knew little about wind power but did know that emerging clean-energy technology could provide jobs in the state. Slowly and painfully, a utility-deregulation bill had made its way through the state legislature in the 1990s. Finneran hadn't worked on that bill, but he had followed its progress and understood its importance. After September 11, the speaker understood its urgency as well.

Hearing Gordon out, Finneran thought the fact that the newest offshore technology could take advantage of Cape Cod's south-shore wind resources was "neat . . . a wonderful coincidence of events . . . a significant amount of clean, green power, forever and ever." He saw Gordon's project as providing new jobs, and when he learned that the turbines would be five or so miles offshore in front of Ted Kennedy's home, the Speaker chuckled a bit, realizing that the senator's opposition would be somewhat "awkward" in light of his opposing drilling in the Alaskan wildlife refuge. How would Kennedy cope with the obvious contradiction? he wondered. He looked forward to watching events unfold, but publicly stayed far away from the divisive issue, which seemed not to concern him.

As the project progressed through its early stages, Finneran watched as the bitterness grew. When he saw Patrick, a new Democratic legislator, make himself an easy target, the longtime politician groaned. When Patrick's Republican opponent settled on the wind farm as a campaign issue, Finneran wasn't a bit surprised. He might have done the same himself. The stage, he thought, was set for a good, spirited contest.

But when the election ended in a photo finish, Finneran took a second look. That wind project was dragging down one of his party's legislators, a fact that he didn't find at all amusing. He began asking questions about what was going on.

By January 1, 2003, the day legislators were sworn in at the State House, the Speaker considered himself up to speed. On the one hand, he was confident from what McDermott had told him that Patrick was legally the winner. On the other hand, amid all the Ping-Pong of legal filings, a Barnstable judge had called for a new election.

At that point, the contested election took on a larger importance. A battle was brewing among the executive, judicial, and legislative branches of the government, and its source, the Cape Wind project, was no longer the central issue. Finneran decided he had to defend the attack on legislative power. He had Patrick seated as temporary representative without being sworn in and appointed a special committee to determine the winner.

Few journalists doubted the outcome. A sculpted cod floats above the state senate chamber's historic chandelier. The old fish is actually a political weathervane, pointing toward the senate president's chair when the legislature is Democratic and away from the chair when the legislature is Republican. In Massachusetts, it's been many a year since the cod pointed away from the president's chair. In most Massachusetts towns and cities, Democrats outnumber Republicans by overwhelming majorities.

After appointing the committee, Beacon Hill bosses chose to ignore the Barnstable judge's order. Wheatley's demand for a new election was met with blank faces, infuriating the candidate. Relatively new to the Land of Our Puritan Fathers, he was slowly learning how Boston played the game.

After briefly considering the situation, Wheatley decided to begin campaigning anyway. If Boston could disregard him, he could disregard Boston. He set up a campaign headquarters, held fund-raisers, and handed out campaign literature.

Meanwhile, as Cape Cod found itself embroiled once more in a Patrick-Wheatley contest, Boston was in a Rip Van Winkle snooze. Eventually, the special committee held a pro forma hearing. Wheatley, in his best suit, tie, and American flag lapel pin, claimed the mantle of voters' rights. Patrick, having been advised to be quiet, smiled a lot and said little. Things continued to drift. Patrick continued to carry out his duties as representative. Wheatley fumed.

Finally, in mid-March, about four months after the election, Finneran's special committee proclaimed Patrick the winner. The next day, Wheatley lost a court battle to nullify the election. The day after that, the House voted to seat Patrick officially.

Still Wheatley pursued the matter. Voters were losing patience. Wheatley seemed to have lost, fair and square. Shouldn't he acknowledge that loss and move on? After all, in the interests of democracy, Al Gore had withdrawn. These court filings seemed like revenge.

Then Mitt Romney, newly elected governor, stepped in, raising many an eyebrow. Patrick's swearing-in ceremony was scheduled for April 2, but at the last minute Romney refused to carry out the duty.

Reporters covered the story in depth. Cape Cod was no longer amused. "As the senior Republican office holder in Massachusetts," wrote Frederick Claussen, Cape Cod's registrar of probate, "I find it troubling to see our Republican candidate for the House of Representatives, coming essentially out of nowhere, to be seen by many as a poor loser, prolonging this controversy as 'sour grapes.' Matthew Patrick won the election—and increased his lead in the recount. At that time, Larry Wheatley should have graciously conceded, and would have gained support and perhaps increased his chances next time around. Not so now." Coming as it did from an important local Republican leader, the letter heartened Patrick.

Still, Wheatley continued to file motions, and the Massachusetts Supreme Judicial Court heard the matter in June. In a relatively short hearing, the justices and lawyers debated details of Massachusetts election law, trying to determine the precise point at which power passed from the courts to the legislature.

In August, the court found in favor of Patrick. Governor Mitt Romney swore in the young representative for his second term on August 27,

2003. "Democracy won out," the relieved state representative told the press on the way out.

Wheatley ran against Matt Patrick again in 2004. He lost by 1,509 votes. In the 2006 election, Patrick was unopposed.

CHAPTER SEVEN

Wianno Seniors, and Those Who Sail Them

I think it's just one of the delights of life, of living on Cape Cod, of watching them, those Seniors race, and also of being at the helm of one.

Senator Edward M. Kennedy, from *Lady of the Sound*, Pearl River Productions

Unable to make Jim Gordon go away, the Kennedy family decided to take their case to the nation. "I would not put a wind farm in the Boston Common or in Central Park or in Yosemite!" shouted Robert F. Kennedy Jr. "There are certain places you should not put it!"

The panic in his voice was disturbing. It was October of 2002, and Kennedy, a well-known environmental leader, was making his national radio debut as a Cape Wind opponent on the National Public Radio show, *The Connection*. A month earlier Kennedy had gone public first on CNN and then in print, when he was quoted in *Time* magazine as saying about the Nantucket Sound shoreline: "People want to look out and see the same sight the Pilgrims saw." While it is true that many people in eastern Massachusetts take the Pilgrim thing very seriously, Kennedy's statement seemed desperate. How many yachts, diesel-powered motorboats, and air-conditioned homes had filled Nantucket Sound's vistas in 1620? skeptics wondered.

Slogans like Kennedy's might work as attention-getters, but listening to someone speak for an hour is quite different from being treated to a sound bite or two. Producers for *The Connection* thought Cape Wind's ambitious project seemed like a story with large national implications. They had heard about the controversy and were familiar with the talking points, but they wanted to know about the substance and science of Gordon's proposal and of opponents' arguments.

Producers set up the story with Jim Gordon in the station's Boston studio. Kennedy spoke on a telephone from New York. Moderator Dick Gordon (no relation to Jim Gordon) felt he knew much of what Cape Wind's Jim Gordon would say, but the mystery for him was why Robert Kennedy, an outspoken opponent of coal and oil interests, would oppose the wind farm. Since Kennedy spoke frequently about the consequences of global warming, his opposition to Cape Wind seemed not to make sense. Were there some important environmental concerns that had gone unreported, the moderator wondered, or did the issue simply come down to a turf battle over who owned the winds of Nantucket Sound?

For nearly an hour on national airwaves, people who had never heard of Nantucket Sound and who knew only vaguely where Cape Cod was located got the full impact of the emotional battle. Jim Gordon and Bobby Kennedy duked it out through the hour-long program, finding absolutely no common ground. Indeed, as the hour proceeded, the energy entrepreneur felt increasingly as though he'd slid down the rabbit hole. The tumble was dizzying.

The interaction was so negative that Greenpeace's Kert Davies, interviewed briefly on the show, chuckled, "I welcome the chance for Greenpeace to be the voice of reason for a change."

Dick Gordon is a smooth referee who can usually handle these crises, but this time he lost control of the conversation. Asked a question, Kennedy rambled on and on, his voice becoming increasingly stressed and his claims ever more surreal. "Horseshoe Shoal is one of the richest fisheries that is close to shore in North America!" Kennedy sputtered.

Hearing the ludicrous statement, which had no foundation in fact, the energy entrepreneur wondered next if the Red Queen would soon appear and shout, "Off with his head!" A highly disciplined man, at times almost military in his demeanor, he managed to keep himself from being drawn into a shouting match, but there were times when listeners could hear his frustration.

Kennedy's "facts" were preposterous and his level of discourse disgraceful, but underneath all that, listeners could hear that RFK Jr.'s grief was genuine. The problem was, the statements he made were just silly.

He talked about blue-collar workers who used the Sound for recreational fishing and claimed they would have to go to New Jersey instead, once the "industrial site" had been built. Why he picked New Jersey was never quite clear. (In fact, the ten-year record of offshore wind in Europe had shown that recreational fishing actually improved. A wind turbine's base acts as an artificial reef.)

Much of Kennedy's fulminations verged on the incoherent. "The aesthetics are going to forbid people from going there," he said. "The marina owners, the tourism industry . . . are going to die." He claimed that property taxes for the "less wealthy, less affluent" on Cape Cod would increase because of the wind farm, a claim listeners could make little sense of. Wind-industry-financed studies had shown that property values near wind turbines do not decrease, while a study paid for by the Alliance claimed that Gordon's wind turbines would decrease the value of southern-shoreline homes. But how Kennedy's extrapolation from this alleged drop to charging middle-class property taxes would increase seemed over the top.

While Kennedy motored on, seemingly unable to take a breath, Dick Gordon tried to moderate the discussion. "Take a break here, take a break," the moderator repeated, but Kennedy refused to comply. At times the show's producers were forced to cut Kennedy off electronically.

For his part, Jim Gordon was caught off-guard by the emotional onslaught. He had known that Kennedy would be on the radio show, but he hadn't realized the show would be done in debate form. As he wondered how to respond to the barrage of statements made by Kennedy, he looked through the glass window separating the interview studio from the head-phoned producers.

They, too, were shaking their heads in amazement. The producers had held Kennedy in high esteem, but after this radio hour, they would never quite see him in the same light.

When Kennedy equated Nantucket Sound with several American national parks, Gordon had had enough. Without shouting, he pointed out that Cape Cod is not Yosemite or Yellowstone Park, but rather a place where about 300,000 people live, many in very large homes that use a lot of electricity. Nantucket Sound was not a wilderness, but a heavily used waterway.

To be sure, Gordon had his own spin on certain issues. The question of aesthetics—what the turbines would look like—had been front and center from the earliest days. Opponents claimed the shoreline would be dominated by looming towers, which was certainly not the case. On the other hand, Gordon said Cape Codders would see only "tiny masts of approximately a half-inch above the horizon."

This wasn't entirely the full picture. Visibility would depend almost entirely on weather. From time to time, more than those tiny masts would be visible. In Europe, offshore turbine visibility varies from month to month, day to day, and even from hour to hour, as weather conditions and the angle of the sun changes. If the day is quite clear and the air very dry and the sun particularly bright and at the perfect angle, the turbine blades stand out against the horizon. And on very rare occasions, when conditions are perfect for the optic phenomenon called "looming," the turbines might appear closer than they really are. But most of the time, summer haze would help to obscure the view from most shoreline homes.

Distressingly for listeners, an emotionally overwrought Kennedy could not be calmed. "If I could just finish this," he shouted when the moderator explained he had to take a station break. Kennedy called Nantucket Sound the "economic driver" for all of New England, apparently forgetting about the high-tech firms ringing Boston, or about the business centers of Hartford or Providence. "This is one of the most heavily utilized recreational boat fisheries in America," he claimed. "He's going to industrialize twenty-eight square miles of the most important tourism area, the biggest tourism resource, in New England."

People who lived in New Hampshire's White Mountains, along the Maine coast, in Western Massachusetts's Berkshire Mountains, or in the Green Mountains of Vermont might have found Kennedy's claim more than a little outlandish.

He continued: "What this project will do, which is what all polluters do, it will make a few people rich by making everybody else poor. It will raise standards of living for Mr. Gordon by lowering quality of life for everybody who lives on Cape Cod, who lives on Martha's Vineyard, who lives on Nantucket, all of them will have a diminishment of their quality of life and a diminishment of their property values."

Later he said to Jim Gordon: "You're a developer here who is trying to make a buck and you're trying to do it by imposing your costs on the public," adding, "I've seen grocery stores on the Cape that go through more of a permit process than this project!"

At this point, it was clear that Kennedy either did not know the basics of the wind-farm discussion or had lost all perspective. The U.S. Army Corps had clearly insisted on a full-scale environmental review, certainly much more in-depth than any grocery store review.

In a sound-bite-driven show, Kennedy's animus might have gone unnoticed. But after an hour, his personal attacks on Gordon were unpleasantly obvious.

Listeners did not like what they heard. The shallowness of Kennedy's rhetoric was easily transparent, so much so that it harmed his case. After the show, listener response was more intense than usual, much of it one-sided criticism of Bobby Kennedy, who listeners heard as aggressive and antagonistic. The show's moderator, Dick Gordon, also finished the broadcast strongly dissatisfied. He had hoped to help elucidate the issues but wondered where the beef was in the project opponents' arguments.

The radio host felt Kennedy relied on sound bites prepared by others, and that, for whatever reason, he had refused to interact with the others on the show, either callers or the project developer. At times, the moderator wondered whether Kennedy was listening to what others said. For someone as articulate as Bobby Kennedy on energy and the environment, thought Dick Gordon, he did an extremely bad job of making his case. The radio moderator regretted his inability to convince Kennedy to come into the radio studio. Perhaps, seated across the table from Kennedy and looking into the whites of his eyes, thought the moderator, he might have done a better job of keeping Kennedy on track.

Other than as a televised presence, Bobby Kennedy was a complete stranger to Jim Gordon. A child of the 1960s, Gordon had, like so many others, grown up idolizing the Kennedy family, particularly after the traumatic assassinations of Robert F. Kennedy Sr. and of the president.

Gordon decided to try again. He belonged to a group known as "E2," for "environmental entrepreneurs," at which Bobby Kennedy was scheduled to speak in late January. Maybe a face-to-face hello with Kennedy might remove some of the bitterness, Gordon thought.

The day was bitingly cold, but more than sixty people, a good number for the new organization, came to hear Kennedy speak. The room was crowded, but Gordon made sure to introduce himself beforehand. Kennedy had seemed cordial enough while shaking hands. Then Gordon heard his project brought up during Kennedy's speech. The environmentalist quickly resorted to demeaning statements, calling Gordon a "thief," an entrepreneur who was trying to "steal the commons" of Nantucket Sound and use it for his own profit.

The room became quiet. Many in the audience knew the Cape Wind team personally. Some wondered how Gordon would react to this out-of-the-blue personal attack.

Gordon's lawyer, Dennis Duffy, figured the statements were just courtroom theatrics, but Gordon steamed. A wind-energy business colleague, Nick Humber, was killed on September 11 while flying to California to discuss Cape Wind issues with wind-turbine producers. To Gordon, building the wind facility was very serious business. To be directly accused of dishonorable intentions by someone as well known as Bobby Kennedy, with no opportunity to respond, was deeply offensive to Gordon.

His childhood working in his father's store had taught him to accept all manner of people. He was at times highly emotional and on occasion given to abruptness, but name-calling just wasn't his style. Listening to Kennedy's freely given personal insults, and realizing that Kennedy would never offer the courtesy of hearing Gordon answer, the energy entrepreneur felt deeply angry.

Cape Wind attorney Dennis Duffy found Kennedy's performance disconcerting. Like Gordon, he idolized the Kennedy family. As a six-year-old, Duffy's mother had watched Ku Klux Klan members parade along Rhode Island streets in 1928, threatening Irish Catholic residents. The Duffy family was one of the first Irish Catholic families to move into a single-family home in Cranston, next to Providence, and neighbors were not happy with their presence there. Duffy's mother would never forget her terror at seeing groups of men in hoods and robes carrying torches in front of her house. When John F. Kennedy won the presidency in 1960, the Duffy family celebrated, believing that the new president had the power to change their lives. When the president was assassinated, Duffy's mother put his picture up on her living room wall, where it stayed enshrined for many years.

Upset to be in conflict with a Kennedy, Duffy hoped to find a way to ease the situation. While Gordon fumed, Duffy walked up after the speech to introduce himself. "I'm one of the Cape Wind guys," Duffy said to the other attorney. Kennedy chuckled, perhaps, thought Duffy, a bit nervously. It's one thing to call someone names from behind a podium, and quite another to speak directly to that person.

"Oh man, I can't believe you think you're going to do this," Kennedy told Duffy. "I learned to sail out there. You're going to have a fight every step of the way."

In the Kennedy family, sailing was not a lighthearted activity. You do business with anyone, but you only sail with gentlemen, J. P. Morgan once said. Old Joe Kennedy intended to make sure that his boys sailed with the gentlemen.

Sailing races were serious business. Once, when RFK Sr. and a college friend joked at the Hyannisport family dinner table about losing a race, the clan's patriarch said huffily that losing a boat race was no laughing matter, then left the dinner table. At old Joe's funeral, Ted told this story: "If we were racing a sailboat, he was there in his cruiser. One time we did badly. He felt it was because we were not paying attention. There was absolute silence at dinner that evening." Second place was not good enough. This created what one biographer called "an aggressive brand of manhood." Old Joe called it "moxie."

This moxie was highly developed in all the Kennedy boys, but particularly in Bobby Kennedy Sr. When four years old, RFK Sr. was teased by his older brothers because he could not swim. Finally, he'd had enough and jumped off the boat into Nantucket Sound, trying to swim. Had his brother not fished him out, he would have drowned. As an adult, he liked to plunge into Nantucket Sound in March, when most Cape Codders are still huddled under wintertime blankets.

Central to this water-dominated machismo was an odd little sailboat, part yacht and in some ways part barge, and rarely sailed anywhere but on Nantucket Sound. Only about twenty-five feet long with an eight-foot beam, the craft was not at all elegantly outfitted but enjoyed a cult following that implied a certain something about its owner. It was the chariot of choice for the wealthy class.

The Wianno Senior, a product of the loins of Nantucket Sound, was an in-crowd emblem—a kind of Social Register, a measure of status, a relic of the past, and symbol of what the world had once been, an emblem of longtime family connections and proof of skill. "Rugged is the word for Wianno Seniors, and Those Who Sail Them," announced a newspaper headline decades ago.

In nearly a century, only slightly more than 200 Wianno Seniors have been built. By 1913 racing had become popular along the southern shoreline, but Nantucket Sound competitors faced several obstacles. The shallow shoals, essentially underwater sand dunes, shifted constantly, so that particularly in spring, an unsuspecting skipper might find himself hove to on a shoal that had moved or grown higher over the past winter. The sound's short and choppy seas also made racing difficult.

In 1913, a group of Wianno families—now dubbed "the thirteen founding families"—asked a builder to design a sailboat to overcome these obstacles. They wanted a short craft with only one mast, with a very shallow draft so the boat could plow through the choppy seas without capsizing.

The vessel they got was a peculiar-looking thing, built for speed and not for comfort. Odd as it looked, though, it met the summer crowd's racing needs quite well. Sitting low in the water, the Wianno Senior was a very "wet" boat, meaning that those who sailed it were unlikely to finish the race in dry clothes. Its 600-pound iron keel kept it upright in rough seas, and its draft of two-and-a-half feet with the centerboard up allowed it to sail over most shoals, including Horseshoe Shoal.

In the 1920s and 1930s, the ceremony of racing the Wianno Senior was very much about maintaining standards. Those watching the race arrived at the long Wianno Club pier at the end of Wianno Avenue—only a stone's throw from where Doug Yearley lives today—dressed for the occasion. The women wore long skirts, large-brimmed hats, and white gloves; the men, impeccable blue blazers, white flannels, and commodore caps.

Much admired were the Parlett sisters, Mary and Edna, who each owned their own Wianno Senior and who each had a man, politely called "the professional," who was responsible for the day-to-day care and sailing of the boat. At race time, the professional sailed the sisters' boats to the starting line and prepared for the race, after which a shore launch brought the sisters out. Mary and Edna, dressed in their finery, graciously stepped down from the launch onto the decks of their respective Wianno Seniors to begin the race. The professional's role was like that of the royal groom who saddled and held the horse for the well-attended Queen of England.

Upon arriving in Hyannisport, Joe Kennedy bought his boys Wianno Seniors. Joe Jr. and Jack took to racing like ducks to water, but Ted seems not to have been quite as skilled. There is a touching photo of a teenaged Ted, dressed in jeans and striped T-shirt, standing at the mast of Jack's boat, learning from his older brother how to raise the sail.

The boat came to symbolize the Kennedy clan. Ted Kennedy went to Chappaquiddick that fateful July 1969 weekend to race Jack's Senior, *Victura,* in the Edgartown regatta. When Ted's daughter, Kara, married Washington architect and champion sailor Michael Allen in 1990, a plastic miniature of the *Victura* topped their wedding cake. The real *Victura* now rests in permanent dry dock, enshrined outside the John F. Kennedy Library in Boston. Another Wianno Senior, RFK Sr.'s *Resolute,* was sailed by his many sons until it burned in a disastrous Osterville boatyard fire.

The Wianno Senior made for great photo opportunities. Joe Kennedy made sure that Jack and Jackie were often photographed on Nantucket Sound in the *Victura*, believing that the carefree life of sailing a small vessel projected an image of his son as young, virile, and ready for action. The images from Nantucket Sound also reminded Americans of Jack's heroic World War II actions, made famous by the Kennedy-sponsored film, *PT 109*. Joe cultivated close connections to the magazine world, and one of the most famous Wianno Senior photos ran on a *Life* magazine front in July 1953. Americans fell in love with Jackie's glowing smile and her elegantly windblown hair. The future First Lady sat on the foredeck of Jack's Senior with her barefoot husband beside her. He looked more than competent to manage his vessel, or, by implication, the ship of state.

Childhood memories can be powerful motivators. The power of nostalgia, particularly for a family that has suffered great losses, is often compelling.

In the summer of 2003, Jim Liedell, a retired utility executive and avid wind-farm supporter, saw Ted Kennedy at a summer concert in Hyannis. Liedell, not being shy, approached the senator. "Can you tell me why you're opposed to this wind farm?" Liedell asked.

"The developer's not paying anything," the senator replied.

"They've said that they're willing to," said Liedell.

"That's peanuts. I'm talking about real money," replied Ted, who in fact had no idea how much the developer would pay, as that detail had yet to be worked out.

"What are your other main reasons?" Liedell asked, persisting with the old silverback.

The *Victura*, the Wianno Senior sailboat owned by slain president John F. Kennedy, is on display outside the John F. Kennedy Presidential Library.

"The sight of them bothers me," said Kennedy.

That's refreshingly honest, Liedell thought, who had grown tired of hearing all the overblown sound bites. Not realizing Kennedy's connection to sailing, the retired executive told the senator that, most of the time, the turbines would be either invisible or barely visible from the senator's home.

"But don't you realize," Kennedy said indignantly, "that's where I sail."

Liedell thought the senator seemed perfectly serious. He also thought that Doug Yearley had, months earlier, been correct. If Nantucket Sound's sailing crowd honestly admitted that their opposition to

Cape Wind was based on their sailing hobby, the complaint might not carry much weight with the American public.

Beginning early in 2002, U.S. Senator Edward M. Kennedy worked the phones hard, calling colleagues, contacting local officials, letting lobbyists know how strongly he felt about stopping Cape Wind. At one point, Kennedy left a briefing on Iraq to ream out a General Electric lobbyist after GE and Cape Wind held a joint press conference to announce the development of a larger offshore turbine, which Gordon said would allow him to reduce his proposal of 170 turbines to 130, while increasing their output.

Kennedy stopped short of publicly opposing the project and offending his labor constituency, who were wholeheartedly supporting offshore wind because of the potential of high-paying jobs, but behind the scenes he threw himself into the job of halting the permitting process. Kennedy aides worried that the senator's position opened him to charges of NIMBYism. Kennedy's friends believed he was digging himself into an emotional hole that would, eventually, do him much political harm. Conversations like the one he'd had with Liedell were bound to get back to the public.

The senator had built himself a powerful position in the Senate and in Massachusetts. Most voters in the state accepted Kennedy's presence as part of the political landscape. Every six years, Republicans dutifully trotted out a candidate who, everyone knew, was merely a lamb brought to the slaughter. No candidate could bring down Ted Kennedy, according to accepted wisdom.

But a few astute observers wondered if that were still true. The Kennedy name had less allure with younger voters, who had not lived through the 1960s and been traumatized by the televised assassinations of the senator's two older brothers. To them, the deaths were just historical events, and Camelot was a movie. They were less willing to allow the Kennedy clan the kinds of privileges that older voters seemed to grant. Indeed, when an older state Democratic Party leader, Phil Johnston, told the *Boston Globe* the nation ought to protect Nantucket Sound out of respect for Ted Kennedy, who had sacrificed so much for his country, many in the younger crowd jeered. Instead, this generation worried about the future of energy in America, and when they studied history and read about a time when gas cost around sixty or seventy cents a gallon, they felt incredulous.

Of all age groups, those in their twenties and early thirties supported Cape Wind most strongly. Kennedy's persistence in opposing the project, many worried, could, at the very best, tarnish his reputation with this crowd. At worst, his opposition could end his hold on the Democratic Party in Massachusetts—unthinkable only a few years earlier.

Kennedy's opposition also made life very difficult for Massachusetts's junior senator, John Kerry. Gordon supported Kerry during his presidential run by holding at least one fund-raiser for him. After that, rumors spread that Kerry actually supported the project. No evidence ever surfaced about Kerry's real opinion, either pro or con, and Kerry in fact tried to avoid the issue. When asked by reporters, he said he wanted to wait for the environmental impact study before deciding, but unlike Matt Patrick, he did not say he would support the project if Cape Wind were given a clean bill of environmental health. Massachusetts renewable-energy fans came to believe that this wishy-washy attitude damaged his run for the presidency, but that seems doubtful. By election time, Cape Wind had made a few national headlines, but it wouldn't become a big-time cause célèbre for many more months.

Meanwhile, Massachusetts's senior senator tried to maintain the illusion that he was neutral. Kennedy aides showed up at various meetings, both public and private, reminding regulators of the powerful senator's interest. But the senator himself, for the time being, kept publicly mum about his intention to stop Cape Wind.

To avoid charges of self-interest or cronyism, the senator needed a public-policy issue. And indeed there was such an issue. Both sides, advocates and opponents alike, agreed that one important question needed to be resolved. Cape Wind, and all other offshore projects built in federal waters, needed a managing agency that would provide oversight and collect lease fees.

For many Cape Codders, the biggest shock when Jim Gordon proposed his project was that, to them, the area of Nantucket Sound seemed to belong to Cape Cod. Legally, of course, they understood that it didn't. But for centuries Cape Codders had fished and sailed in these waters. How could a private developer suddenly show up and take over such a large area of what seemed to them to be public land?

Their confusion was understandable, seen in the context of New England history. Elsewhere in the country, private businessmen like ranchers routinely pay lease fees to use public lands. Gulf of Mexico shoreline residents long ago adjusted to the sight of offshore oil and gas rigs. In Pennsylvania, when a coal-powered generating plant is proposed near a small village, local people know they have little recourse. And in the American West, homeowners must allow drilling rigs access to their private property when, as often happens, surface rights and mineral rights are separately owned.

In New England, however, property titles sometimes reach all the way back to the days of the kings George. England and Scotland had enacted land-conservation and wildlife-management laws long before the Pilgrims, who carried these laws with them, founded Plymouth Plantation. Several centuries of comparative stability buttressed by democratic institutions like Town Meeting mean that New Englanders presume a much greater right of involvement in local matters. In Pennsylvania, locals take for granted that a power plant might be sited immediately abutting their property; New Englanders have a different point of view entirely.

The region has consequently escaped many of the modern realities taken for granted elsewhere, so that a highly visible offshore wind farm could indeed be shocking. "How can they do that?" an astounded New Englander might ask about a wind farm proposed for federal waters. But if he asked a Gulf Coast resident, that person might not even understand the question. In his world, such projects are common.

Nevertheless, most people agreed that some sort of federal oversight agency would have to be responsible for offshore wind development. Most parties, except for the Alliance, agreed that the U.S. Army Corps of Engineers had the right to "permit" the project, meaning they could examine environmental impacts and specify construction details. If the project were *not* appropriate, the Corps could say so, or it could list so many conditions of construction that the developer could decide to walk away.

But the Corps did not have legal jurisdiction to oversee the project once it was built. It did not have jurisdiction to collect lease fees, and, perhaps most important, it did not have jurisdiction to undertake a wide-scale study of where offshore wind projects might be built. Some parties believed that the siting of offshore wind projects ought to be planned out by a central agency, while others believed that such planning ought to be done loosely, if at all.

But all parties, including the Corps, agreed that the jurisdiction provided the Corps by the 1899 Rivers and Harbors Act was too limited in scope. Indeed, in the summer of 2002, Colonel Thomas Koning, the New England regional head of the Corps, went on record with his concerns, specifically asking a federal commission on the future of ocean policy for "a national policy for non-extraction ocean energy projects and commercial ventures."

There was a federal agency that did have experience in offshore leasing. The Minerals Management Service (MMS), a bureau within the

U.S. Department of the Interior, had for years overseen offshore oil and gas leasing. As 2002 progressed, many people began discussing whether MMS would be the appropriate agency to oversee future proposals.

Kennedy, however, wasn't looking to the future as much as he was looking at the project that had already been proposed. Planning for future offshore wind was fine, but Kennedy wanted to stop the proposal currently on the drawing board, Cape Wind. Massachusetts's senior senator put forward the first of what would become a string of increasingly aggressive legislative attempts to block Gordon. He began floating some language to colleagues that proposed a two-year National Academy of Sciences study of offshore wind development on the Outer Continental Shelf, the relatively shallow undersea area of the Atlantic that runs up and down the nation's East Coast. When the language surfaced, Kennedy and his staff gave contradicting statements as to whether or not this proposed study carried with it a moratorium on offshore wind, which would, of course, involve Cape Wind. "This is not a delay mechanism," said Kennedy aide Stephanie Cutter. On the other hand, the *Cape Cod Times* reported that Kennedy did intend the legislation to include a two-year moratorium. "Before moving forward, I think it's important to wait for the results of the study, which will shed significant light on the impacts of the project," Kennedy reportedly told the *Cape Cod Times.*

Most people agreed that some sort of planning for the future ought to occur, but the two sides were bitterly split over whether that planning should involve the Cape Wind proposal. Project supporters claimed it was simply a delay mechanism. Renewable-energy supporters saw it as delaying the fruition of an important new strategy to develop clean energy. Unions saw the language as a way to delay the development of an industry that promised high-paying jobs. The business and investor community saw it as an attack on free-market principles, since Gordon and his investors had spent money in good faith, based on the legal ability of the Army Corps to permit the process. Phone calls flooded Kennedy's offices.

Throughout the first half of 2002, various versions of the Kennedy language surfaced, sank, and resurfaced. These forays found little traction. Indeed, Kennedy's attempts were backfiring, by forging some hitherto very unlikely alliances. Prior to the Cape Wind proposal, for example, unions and environmental advocates had rarely seen eye to eye. Now they were new best buddies.

Meanwhile, the offshore oil and gas industry was looking into the future of renewable energy at sea. The industry had long sought a solution to the problem of their abandoned offshore drilling platforms, which federal law required be removed when they were no longer in use. Claiming the removal could cost $10 billion, the industry investigated converting the platforms into bases for solar arrays or wind turbines. They, too, needed an oversight agency.

The most obvious choice was the Minerals Management Service, which the oil and gas industry was already accustomed to working with. It was the Minerals Management Service that had permitted the industry's drilling platforms and oversaw the industry's offshore operations. Wyoming Republican Congresswoman Barbara Cubin, said to be carrying water for the oil and gas industry, submitted authorizing legislation. Hearings on Cubin's proposal occurred during the summer of 2002, but the legislation did not move forward. Washington was preoccupied by the lead-up to the Iraq war.

Legislative finagling returned to the Cape Wind forefront in January 2003. This time the targeting of Cape Wind was much more blatant. Massachusetts Senator Ted Kennedy and Virginia Senator John Warner, Bunny Mellon's former son-in-law, were rumored to be shopping around a rider to an omnibus appropriations bill that would deny funds to the Army Corps for use in permitting Cape Wind. Specific language never turned up, but Kennedy did admit to being interested in the issue. "The senator is open to legislative language to delay this wind-farm proposal from moving forward until Congress has time to consider a project of this magnitude that has such huge ramifications, including benefits to the environment," a Kennedy aide said.

Each legislative delay tactic created more support for Jim Gordon, as popular resentment against backroom legislation grew among voters, journalists, and other elected officials, who, in light of the war in the Middle East, found Kennedy's use of Congress to achieve a private goal to be embarrassing. In particular, Massachusetts's state senators and representatives, who had passed the state's electricity restructuring legislation and who had watched Matt Patrick being dragged through a legal meat grinder, had had enough. They wrote several letters reprimanding their Washington colleagues. "We are writing to oppose any attempt to legislatively block or delay the ongoing regulatory review of the Cape Wind project," said one letter. "As we understand it, there have been discussions regarding a Senate appropriations rider that would, without a hearing on the merits" end the Army Corps' review of Cape Wind. "If parties have substantive concerns with this or any other renewable energy project, it would be more appropriate to address such concerns" in public, and within the context of the Corps' permitting process.

If Kennedy was trying to keep his head down, he wasn't doing a very good job of it. At bottom, friends knew, the senator's deep emotional connection to Nantucket Sound was the real issue. In 1971 Kennedy had filed a bill that would have enlarged the Cape Cod National Seashore by including the islands of Nantucket and Martha's Vineyard. In 1972, when islanders responded bitterly to that proposal, he filed the Nantucket Sound Islands Trust bill, to preserve as "forever wild" large portions of the islands in Nantucket Sound. Some lands would be cataloged as "scenic preservation" areas that would have strictly controlled building, while other regions would have relatively loose construction controls. U.S. Congressman Gerry Studds, Mark Forest's former boss, filed a companion bill in the U.S. House of Representatives.

Bearing the taint of condescension, in the ancient and royal sense, the Kennedy bill pitted local year-round residents against the wealthy summer folk by creating a unique federal commission that would have authority over the islands and would be composed of both groups. Kennedy and Studds saw this as a way to circumvent small-town politics in order to protect the region from overdevelopment, but many local folk resented the intrusion. "All the Kennedy bill does is take away people's rights to their property," said Martha's Vineyard politician Robert J. Carroll. Another local politician, Herb Hancock, called Kennedy's bill "a federal land grab" meant to benefit only "raccoons and conservationists." Many year-round residents believed the commission would increase the power of the wealthy crowd who saw Nantucket Sound as a summer park—the population increases by as much as 500 percent in the summertime—rather than as a place where people had to live, work, and pay bills.

Kennedy's bill had cosponsors from Pennsylvania and from Colorado, but it did not have the sponsorship of his cosenator, Republican Edward Brooke, who in fact owned some Martha's Vineyard property. "I am not yet convinced that it is necessary for the federal government to intrude so pervasively in the lives of my fellow islanders," Brooke said. A Vineyard referendum turned down Kennedy's idea by 60 percent.

Boston television personality Natalie Jacobson once asked the senator to name a personal strength. "Perseverance," he answered. That was an understatement. In the halls of Congress, Kennedy was known as a bulldog. When Connecticut senator Chris Dodd failed on a first attempt to get a bill passed, Kennedy told him it was "not uncommon" for legislation to take five or six years to pass.

Like Jim Gordon, Ted Kennedy did not know how to walk away from a fight. Kennedy and Studds continued their efforts into 1976, when compromise language, very much watered down from Kennedy's original idea, finally was passed. The language created several regional planning commissions that would review large-scale proposed developments and have some very limited authority over those developments.

Through the five-year battle, year-round folk remained very bitter, while island celebrities generally supported the senator. Among them were Mia Farrow, Andre Previn, Jules Feiffer—and Walter Cronkite.

CHAPTER EIGHT

The Fourth Estate

Freedom of the press is guaranteed only to those who own one . . . .

A. J. Liebling

Walter Cronkite had long been connected to the Kennedy family. Ernie Corrigan, the Alliance's public-relations man, once bragged to Jim Gordon that acquiring Walter Cronkite was his own little "achievement," that he had met with Cronkite and convinced him to support the Alliance and to oppose Cape Wind. But people who knew Cronkite realized there was much more to the story than just a meeting or two with a public-relations consultant.

For much of his life, Walter Cronkite had a long and comfortable relationship with the Kennedy family, and was on a first-name drop-by-for-dinner basis with many members of the family. "Occasionally we have sailed from our Martha's Vineyard summer home across Nantucket Sound to Hyannis," the newsman wrote in his autobiography, a book about his close contact with a number of powerful families, rather disingenuously entitled *A Reporter's Life*. "We drop anchor near the

Kennedy compound, and almost always a vivacious and gracious Ethel Kennedy drops by in her sailboat. She insists that we come ashore, play tennis if there is time, and stay for dinner . . . ."

As far back as the 1920s, old Joe cultivated close relationships with newspaper, radio, and magazine journalists. As television gained momentum, the clan made it a point to do the same with on-screen personalities, ensuring that as Americans switched from newspapers to television to get their daily dose of news, the photogenic Kennedys became symbols of a new national vitality.

Walter Cronkite was there to chronicle it all. When Jack ran for president in 1960, Cronkite asked the unaskable on TV—whether Americans would elect a Catholic to the presidency. The question incurred the clan's wrath, but their anger wasn't strong enough to cause them to cut off relations with the powerful newsman. Shortly after that interview, Cronkite did a cozy little at-home-with-Jack-and-Jackie piece, to be paired with a similar interview with Richard Nixon and wife. After the election, Cronkite chronicled the major events of Camelot, like the Bay of Pigs, the Cuban Missile Crisis, the Berlin airlift, and Jack and Jackie's artistic events.

After the Kennedy assassination, Cronkite kept in contact with the family, attending Ethel and Bobby Sr.'s raucous parties at Hickory Hill, where high officials, famous academics, actors and actresses, foreign revolutionaries, and foreign dignitaries might be pushed fully dressed into the swimming pool. By the time of the Tet offensive in Vietnam, Bobby Sr. and Cronkite were on a cordial basis. In 1968, Bobby Sr. called Cronkite over to his office. "When he expressed his strong belief in the necessity of extricating ourselves from Vietnam," Cronkite wrote, "I gratuitously suggested that he ought to take his argument to the people by entering the presidential primaries that spring." Bobby countered that compliment by telling Cronkite that "we want you to run for the Senate this year." Bobby would be leaving his post as a New York senator, and the family thought Cronkite would be a good stand-in.

Shortly after that, Bobby Sr. used Cronkite's television show, *The Cronkite Report,* to announce his long-anticipated run for the presidency, putting the debate about whether he would run or not to rest. Some biographers say that even his closest family members, like Teddy, hadn't known for sure whether Bobby would run until they saw Cronkite's show.

As the years rolled by, Cronkite, like the Kennedys, assumed an iconic celebrity. As a kid, Cronkite had helped the mother of a young friend pick out a toy sailboat as a birthday present for her son. In his book he says he was a bit jealous, and when he grew up he bought his own sailboat. And then another and another. Eventually both the boats and his friendship with the wealthy and powerful increased in scope and brought him to Nantucket Sound. He bought a house on Martha's Vineyard. In his memoir, Cronkite writes that his wife complained: "'Doesn't anybody ever buy a smaller boat?' The answer is: Not if they can help it."

So it surprised no one when Cronkite appeared with RFK Jr. on the September 2002 CNN piece about opposition to Cape Wind. What *did* surprise some locals was that the newsman was so outspokenly hostile to the project. The scientific review process had barely begun, yet Cronkite was offering his iconic celebrity to the opposition crowd. Although Bobby Kennedy Jr. had also been featured in the CNN piece, Cronkite was obviously central, shown sailing his large yacht in Nantucket Sound and sitting for an interview, dressed casually in his pleasant Martha's Vineyard seaside home with large sailboats parading by outside the window—a scene of wealth, power, and confidence.

The piece began with Gordon introducing his project by explaining that it would encompass 170 turbines spaced one-half to one-third of a mile apart. It quickly jumped to Cronkite. "That's right in the middle of

Nantucket Sound!" he told the cameras. "When they fill it up with these things, it's going to be disastrous."

"Are these just ugly things to you or are they conceivably part of a landscape that we might just get used to?" CNN reporter Michael Shoulder asked.

"No, I don't think these things can possibly be considered attractive in any possible way. I don't care what colors you paint them or whether you have them dance in unison to music or what. They are big ugly things sitting out there in the middle of what should be the pristine waters. The way we are affected in a visual sense will be nothing compared to natural life—how it is affected out there."

"Do the whales know how to get around the towers? Do dolphins know how to get around those towers? I don't know. We'll have to find that out but it sounds to me like they are going to have a very tough time."

With that statement, whale-mongering reached its peak. At that time, a body of solid science had been established in Europe that showed that whales, seals, dolphins, and the like left the area while turbine construction occurred, but later returned. There existed absolutely no scientific evidence of long-term harm to whales, but animal rights groups pushed the issue very intensively anyway.

For Gordon and Cape Wind staff, Cronkite's entry into the fray was discouraging. "If I've lost Walter Cronkite . . ." Lyndon Johnson moaned after watching Cronkite proclaim Vietnam unwinnable. Watching Cronkite deride his project, Gordon understood the power of the voice and image of the iconic old newsman. The project's wealthy foes had assembled a truly formidable team.

Cape Wind staff had had a horrendous time with the local press. The small community daily newspaper, the *Cape Cod Times,* with its publisher who lived on Wianno Avenue and its editor who told people he hoped to win a Pulitzer by uncovering the truth about Jim Gordon, was supposed to be acting as the "paper of record" in reporting Cape Wind's development. This meant that other journalists outside of Cape Cod took the paper-of-record reporting at face value. As the months passed, those who followed the paper's coverage of the wind farm wondered about the personal nature of the paper's attacks on Gordon.

In January 2002, the paper began writing anti-wind-farm editorials that were sometimes thousands of words in length. The first, "Wrong Place for a Wind Farm," said the newspaper had done "a serious examination" of the wind proposal, a claim that irritated Gordon, who had yet to finalize the project's details. "This industrial park on the water will hurt our environment as it affects sea bed fauna and flora, mammals, fish and migratory birds," the editorial said without backing up its claims. It ended with a call for the Cape Wind project to be sited—somewhere else. The editors suggested Massachusetts's Fall River, a poor city full of immigrants and people of color where a highly polluting coal plant already existed. Put it there, the editors wrote. Or maybe in Delaware. Or maybe in New Jersey . . . . The irony—that Cape Cod's environment needed to be protected, but that Delaware's or New Jersey's were fair game—was not lost on many readers.

The paper's news editor, Paul Pronovost, sometimes refused to print corrections to factual errors. At one point a *Times* editorial copied, almost verbatim, several paragraphs of a letter signed by Doug Yearley, without using quotes, so that readers had no way of knowing they were reading Alliance literature rather than the writing of the *Cape Cod Times* editorial board.

Perhaps even more worrisome were the constant "news" stories. Fanning the flames of hysteria, the paper often ran screaming headlines across the top of the front page, an onslaught that had a powerful effect on readers. One headline coupled Jim Gordon with the energy giant Enron, which was later followed by an editorial stating that Gordon was "partners" with Enron, after which angry project supporters called Cape Wind, asking why the company had deceived them by hiding its affiliation with the energy giant, then in its scandalous death throes. Cape Wind employees had to explain that the company wasn't partners with Enron, but was merely talking with Enron Wind, which manufactured turbines Gordon was considering buying. (Enron Wind was bought by General Electric, and Gordon is still considering buying their turbines.)

In another case, the paper ran at least six stories and editorials on a "company," Sea Energy Generation that supposedly was building another Cape Cod offshore wind-energy project. A reporter from the *Washington Post* picked up that story. "This is like an Oklahoma land grab," complained lobbyist John O'Brien, identifying himself in the *Post* as president of a grassroots group saving Nantucket Sound, but kept quiet about his position as an Alliance lobbyist. In fact, no Sea Energy proposal was ever submitted. A man named Dave Spaulding visited the Army Corps once for a preliminary informational visit, then dropped the matter. Such early fact-finding meetings are common, but most companies bring teams of people and take reams of notes. Spaulding showed up alone. Karen Adams, the Corps' wind-energy point person, noticed that Spaulding took no notes, brought along no paperwork and left without submitting any documents. As of this writing, Spaulding has never submitted any documents, a fact *Cape Cod Times* readers never learned.

One explanation for the skewed coverage lay in the fact that *Times* management were running a "page-one game"—a contest to see which reporters could get the most front-page stories in any given month. The winner received a day off and a gift certificate to a local bookstore. Consequently, when *Times* editor Cliff Schechtman made clear his opposition to the Cape Wind proposal, reporters competed with each other to run "big" stories that "exposed" Jim Gordon and Cape Wind.

Cape Wind executives and project supporters were frantic over what to do about the *Cape Cod Times*'s behavior. Gordon had tried several times to meet with editors, but the few meetings the editors agreed to convene deteriorated almost immediately into attack sessions during which Gordon was ridiculed in many ways. *Times* editors were particularly furious that Gordon would not open his company's books for their inspection—something private companies rarely do, but which *Times* editors implied was an egregious fault.

The worst was yet to come. In January of 2003, the Alliance began airing on local access television a long and expensive infomercial about the "treasure" of Nantucket Sound and the evils of offshore wind energy. Beginning with an invocation of Henry David Thoreau and *Moby-Dick,* the emotional piece called Nantucket Sound a "163-square-mile sanctuary," implying marine sanctuary status, which did not exist. The film depicted playful whales cavorting and diving, although whales had not frequented those waters for decades.

Bookending the misinformation was Walter Cronkite, this time dressed seriously in dark jacket, tie, and white shirt. "I'm very concerned about a private developer's plan to build an industrial energy complex across twenty-four square miles of publicly owned land in the middle of Nantucket Sound," Cronkite said, knitting his eyebrows in earnestness. He reappeared at the end of the piece, saying, "Our national treasures should be off-limits to industrialization."

Unable to get basic information to the public via the local daily newspaper and overwhelmed by the Alliance's "grassroots" blitz, Gordon was running hard, speaking to local groups whenever he could, writing letters to the editor and writing opinion columns, looking for ingenious ways to get his message out. But newspapers outside Cape Cod, not yet alerted to the project's significance, generally back-paged the issue.

Then Gordon got a big break. The wonderfully outrageous Elinor Burkett arrived. At a time when Cape Cod was becoming more and more homogeneously upscale, Burkett clearly came from somewhere else. And not just from anywhere, but from New York, from a seriously funky New York. An independent writer, with the emphasis on "independent," Burkett did not wear pearls-'n'-pink. She would have been noticed immediately at the Wianno Club. She wore an old green L.L. Bean down jacket with brown leather trim, had outrageous hair, and paired a warmly eccentric temperament with a genuinely gleeful smile when provoking people of a certain sort. On assignment to cover the Cape Wind controversy from the *New York Times Magazine,* Burkett arrived on Cape Cod with an expense account in tow, as well as a dog and a husband.

Cape Wind had hit the big time. Jim Gordon welcomed her with open arms. Finally, he thought, a real journalist, someone who will take some time to dig. Initially at least, the Alliance was equally overjoyed. Its then director Isaac Rosen, who for a brief time had been a stringer for the *Boston Globe* and whom Burkett described as "Woody-Allenish," invited her to stay with him in his home while she worked on her story. Burkett declined, but did allow Rosen to cook her dinner. Both sides attributed the magazine writer's wintertime presence in the hinterlands to their own behind-the-scenes actions. Burkett, who calls herself a liberal and Ann Coulter a friend, wasn't saying what had brought the matter to her attention, other than to tell all who asked that she had seen some of the news clips and found the controversy interesting.

The first place Burkett went upon arrival was Cape Wind's communications office. "I'm Ellie Burkett and I'm calling from the *New York Times Magazine,*" she said when she called for an interview. Mark Rodgers paused for a few seconds. Then it sank in. "I've heard of that publication," he said, somewhat lamely. It happened that the magazine was Rodgers' favorite, something he tried to make time for every week because he enjoyed the more in-depth articles he found there.

Rodgers began to explain the nuts and bolts of Cape Wind's proposal. Burkett immediately cut him off. "I already know all that. I've read everything on your Web site, and on the Alliance Web site," she said abruptly, almost spitting out the words with her typical impatience. Rodgers found her direct, straight-to-the-point manner refreshing. He was also impressed that she'd read everything. In his experience, few journalists read much of anything.

Later, Burkett and her husband took Rodgers out to dinner. Rodgers found himself entranced by Burkett's character and impressed by her no-nonsense approach to spin. She made clear to him that she was only interested in facts, and that she intended to speak to every person she could find who had an interest in Cape Wind.

At one point, Burkett dropped by Delahunt's Cape Cod office, where the congressman was, coincidentally, giving another of his anti–Cape Wind tirades disguised as press conferences. Delahunt "seemed vile to me," Burkett said later. "He had this person who worked for him who was unbelievably cocky and arrogant," she said, referring to Forest. "He was totally and blithely convinced that they would get their way. There was no doubt. And there was no doubt in his mind that he was right."

When Burkett asked Delahunt questions during the press conference, she was even more disappointed. "When I asked for the facts, he just had a kind of bluster," she said.

After meeting everyone the Alliance asked her to meet, she remained unimpressed. She ended up deciding that the arguments were "nonsense" and that the nonsense would make her job even harder, because she would have to struggle to try to find a way to be "fair." Eventually she decided that project opponents simply didn't know what they were talking about.

Alliance officials arranged for a telephone interview with Walter Cronkite, an experience she found even more astonishing. She asked why he had made the infomercial, what his objections to the project were, and how he squared his objections to his writings about the threat of global warming and the importance of clean energy. It was a short telephone interview. "He had nothing to say," Burkett recalled.

Cronkite told her he had received mailings and read about the "scheme" in the local newspaper. It "upset me considerably," Cronkite told Burkett. "It will certainly be ruinous to several aspects of life on Nantucket Sound. Early evidence is, there is danger to wildlife." He mentioned the usual icons—whales and porpoises. "It's a marvelous boating area for recreational fishermen and sailors. It would severely interfere with that, of course . . . .It will be most unsightly."

"I'm all for these factories," Cronkite said. Burkett thought he must have made this statement fifteen times. "Wind energy should be used. But there must be areas that are far less valuable for their other attributes than this place is." He suggested that the deserts of California would be a good place, or perhaps the New England mountains. "Inland in New England would be a good idea," he said, repeating the Alliance mantra that Cape Cod waters were so much more special than other places on the earth.

Burkett was losing patience at this point. By now in her midfifties, she had been a reporter for the *Miami Herald,* earned a doctorate in history at the University of Pittsburgh and a master's degree in journalism from Columbia University, authored six books, and traveled around the world. She was demanding, fast-paced, and experienced. She knew a scam when she heard one.

"I'm being brutally honest here," she recalled later. "I thought he was a pitiful old man who was being used . . . . You could tell that he was uncomfortable with the knowledge that what he was doing was morally wrong." Cronkite ended the short telephone discussion with a request that particularly saddened the outspokenly blunt Burkett: "Be kind to an old man," he said.

But Burkett's most scathing views were reserved for *Cape Cod Times* management, who, in her opinion, had violated journalism's most basic ethics. She began looking at the journalism question by lunching with Schechtman. The lunch itself had little content, she thought. It seemed more like a hail-fellow, good-will kind of thing. We're all in this together, and so on. She told him that there didn't seem to be any line between the editorial page and the news in his paper.

By that time, Burkett thought it obvious that anything in favor of the project would have made Schechtman angry. Nothing positive about Gordon is going to show up in this guy's news pages, Burkett thought. "The *Boston Globe* stuff was the reasonable stuff that you would expect from the professionals," she decided later. "The *Cape Cod Times* was whipping people up and telling people they should be hysterical. There was no doubt in my mind that the *Cape Cod Times* was responsible for most of the hysteria."

Burkett thought project opponents had one severe handicap: They desperately wanted to vilify Gordon, but they had nothing. Despite all their research into Gordon's background, Burkett thought, they had found no way to ruin his reputation. Burkett had this sense that it was driving opponents crazy.

Senator Kennedy refused entirely to speak with her. He was, from her point of view, perhaps the worst of all the opponents. After opposing many fossil-fuel endeavors, he was doing everything behind closed doors to stop Cape Wind. His actions may have been hidden, but that his hand was at the helm was all too obvious.

Burkett galloped out of town as fast as she rode in. Everyone was excited. Some were impressed. Some were depressed. A few were greatly amused. She was more fun in the deep, dark dead of winter than most Cape Codders had had in a long time.

Then began the long wait for the magazine piece. For months, Rodgers and others looked at each Sunday's magazine, jazzed to see what Burkett would write. It didn't appear. The Iraq War had intervened. During the editing process, the editors kept saying: This is it? There must be more meat to the opponents' claims, they insisted.

There was no pressure placed on her, but Burkett had to explain that the piece would not contain any kind of line-by-line "balance." She went to great lengths to put in everything project opponents had told her, but the bottom line was there just wasn't much there.

The piece finally appeared in mid-June, in time for the summer season. It was not kind to the Alliance. Burkett's sharp outspokenness slammed the Alliance and quoted Cronkite himself saying, "The problem really is NIMBYism, . . . and it bothers me a great deal that I find myself in that position." She quoted Yearley saying, "I don't know what the rush is" in permitting the project, and Schechtman complaining that Senator Kerry wouldn't chime in, even though he owned a house on Nantucket. Additionally, she quoted many project proponents who universally praised Jim Gordon as highly qualified, the right man to move the nation forward by developing such an innovative project.

Mark Rodgers found the article online about 5 P.M. the Friday before Sunday print publication. Calling Dennis Duffy's cell phone, he got him on a ferry ride over to his family home on Martha's Vineyard and read the whole thing to him over the phone. Both were thrilled, but their reaction was nothing compared to Gordon's. He was overjoyed. Finally, serious national coverage of his project. If it were killed now at least it wouldn't go down in the dead of night without anyone knowing.

For the Alliance in general and for Cronkite in particular, the piece was devastating. The blowback was intense. Once dubbed "the most trusted man in America," a title that he took quite seriously, Cronkite pulled in his horns and retreated to his own private arena.

The infomercial, however, kept on playing.

For Gordon, Burkett's piece was a major turning point, allowing him for the first time to switch hats. On Cape Cod, because of *Times* coverage, he was usually the evil profit-driven developer, but after Burkett's piece, he felt he could wear the white hat he preferred. Ushering in the summer, Burkett's story alerted reporters to the fact that there was a very cool story on Cape Cod, and there probably isn't a television reporter alive who doesn't want to go to Cape Cod on an expense account during the warm months.

Fox News, ABC's *20/20, CBS Sunday Morning*—"It reads like a summer novel," was their opening line—they all showed up. Eventually even Japanese and Korean television crews appeared, entranced by the seemingly inane American argument. Gordon was having a ball. He'd always wanted to be a movie director and now, in a sense, he was.

Cronkite, on the other hand, was taking a beating. His name was appearing under stories with headlines like, "Celebrities Blow Off Wind Project." Despite the cooling breezes of Nantucket Sound, Cronkite was having a long, hot summer.

Meanwhile, under the radar, back-channel diplomacy was under way. George Woodwell, the Woods Hole environmental scientist, was doing what scientists do: He was writing letters. Woodwell had found Cronkite's infomercial a pretty powerful disservice to the public interest.

Writing the newsman, Woodwell asked Cronkite not to think provincially, but to look at the larger issue. The letter included a kind of executive summary of future climate chaos, and added that, in the absence of governmental leadership, entrepreneurial endeavors like Cape Wind were all the country had to rely on. "I hope that you will reconsider your opposition to the wind farm which is, after all, a very small intrusion by comparison with, for example, the Glen Canyon Dam," Woodwell wrote in closing.

Cronkite called Woodwell almost immediately after receiving the letter, which surprised and impressed the scientist. Woodwell felt Cronkite remained very much at arm's length during the conversation.

The scientist followed up the phone conversation with yet another letter, this time responding specifically to Cronkite's concern that Cape Wind was a private, for-profit venture. "It is no more commercial than other power plants that usurp air, water, land in varying degrees," he wrote, adding that Cape Codders were "bathed in effluents" from the infamously polluting local coal plant, Brayton Point, located near Fall River, as well as from coal plants as far away as Ohio. If Cape Wind were built and the nation moved ahead with emerging energy technology, he wrote, "perhaps we could then, in due course, join in a major effort to replace the power produced by Glen Canyon Dam and restore a canyon that should never have been lost."

Woodwell had an impressive scientific background that began around the time Rachel Carson was writing *Silent Spring*. In 1957, Carson received a letter from a friend who lived near Duxbury, just north of Cape Cod. Unbelievable as it is today, during the 1950s, countywide aerial spraying of the insecticide DDT was routine. Marketing videos showed pictures of kids playing during street spraying of the chemical. The Swiss chemist Paul Hermann Muller of Geigy Pharmaceutical, who discovered the chemical's insecticide properties, received a Nobel Prize in 1948 for his work. Carson's friend's letter told a different story. She wrote that after the spraying, birds dropped out of the sky with their claws clenched to their breasts, apparently in some kind of agony.

After years of research, the courageous Carson published *Silent Spring*, which helped launch the modern environmental movement. The work claimed that DDT and other manufactured chemicals were killing wildlife, destroying nature, and even making humans themselves ill. The chemical industrialists, some of whom summered on Oyster Harbors, fought back. First came the usual sexist epithets like "nun of nature" and "hysterical." Next came lawsuits. The barrage hadn't stopped by the time of Carson's death in 1964.

While Carson was writing her book, George Woodwell was looking at DDT's specific effects in a forest ecosystem in Maine and eastern Canada. In 1957, the routine wide-scale spraying of DDT occurred in northern forests to kill the spruce bud worm. Woodwell, a young botanist at the University of Maine, found DDT accumulating on the ground, making its way up the food chain in ever-more-concentrated amounts, eventually killing osprey and other species. Half the DDT never hit the ground, he found. That DDT instead floated on air currents and killed species in far distant places. Initially Woodwell had believed DDT spraying was essential, but his research over ten years confirmed Carson's brilliant analysis based on fragmentary information in her best seller.

The chemistry of the environment is an essential basis for its functioning, he would come to understand. "We can't violate that chemistry in any way, and that goes for fossil-fuel emissions, which change the chemistry of air, water, and land," Woodwell believed. "Industrial society has to be rebuilt so that it doesn't change the basic chemistry and physics of our living environment." As with Carson, chemical manufacturers fought back by threatening lawsuits. Woodwell sued right back, helping found the Environmental Defense Fund. He was an original director of the Natural Resources Defense Council, a position he holds today.

Upon receipt of Woodwell's letter, Cronkite began to feel some doubts. In late August 2003, a Cronkite column outlined his concerns about Cape Wind and reiterated his opposition "until the developers can assure me that the sound is the only feasible site for this regional field of generators . . . ."

After reading that caveat, Mark Rodgers called Gordon, who was over on Nantucket entertaining friends for the weekend. "I think this gives us some leeway," Rodgers said. Gordon called directory assistance, got the newsman's phone number on Martha's Vineyard, and called him. He was astonished when Cronkite himself answered the phone, listened, and invited him to fly over Monday morning. By 9:30 that day, the pair were studying maps and charts of the Sound and of the project.

For Gordon, who couldn't quite get over that he was talking with the real-life version of a television image, the meeting felt surreal. Cronkite asked questions about offshore wind and about Cape Wind specifically. He also asked about Gordon's background, his experience in energy development, and his commitment to emerging energy technology, none of which had appeared in the local newspaper. He told Gordon that his children and grandchildren had already razzed him about opposing Gordon's project. Gordon presented the project's facts, countered the Alliance-perpetrated myths and explained that Cape Wind would likely partially offset the operation of Cape Cod's oil-fired electricity-generating plant. When he learned that the Alliance had not explained to Cronkite the project's extensive environmental-review process, Gordon talked about that also.

Cronkite told Gordon that, in effect, he'd been blindsided by the Alliance. He had not expected to become the opponent's poster boy, he said, then apologized for having opposed the project without first having researched the facts. Gordon left the meeting impressed by the elderly man's honesty.

"Please consider withdrawing," Gordon asked. "Because you are so well respected, because of your influence, some people may not be giving us a fair hearing." The newsman made no promises.

On August 28, Cronkite picked up the phone and called the Associated Press to announce that he no longer opposed the project. He also insisted that the Alliance no longer use his image in their infomercial. The newsman was particularly annoyed because, without his permission, the Alliance had been sending the video out as part of their massive fund-raising package. Exactly how much money the Alliance made by marketing Cronkite's image was never revealed.

In the midst of this Hallowed Halls of High Journalism drama, the profession's working stiffs started sniffing around. Media critic Mark Jurkowitz, then with the *Boston Globe,* started making calls at just about the time Burkett rode into town. Jurkowitz, hailing from northeastern Pennsylvania coal country, graduated from Boston University's School of Communications in 1975. After several years of knocking around out west, he began writing features and news for various Greater Boston newspapers, but his real interest was the media itself. He wrote a column called "Don't Quote Me" for the *Boston Phoenix,* an independent weekly, then began writing for the *Boston Globe.*

Jurkowitz intended to cover the *Cape Cod Times* itself, rather than the project. He had a reputation for being calm, judicious, and measured in his writing, but criticizing a newspaper is never easy. Jurkowitz was intrigued by the role the *Cape Cod Times* was playing in the Cape Wind story and accepted an assignment from a small Boston-based quarterly public-policy magazine to investigate the issue. Editors do not like to answer the phone and find Mark Jurkowitz on the other end, but initially at least, when the media critic called Cliff Schechtman, the editor seemed cordial enough.

To Jurkowitz, the journalistic issue was not whether their editorials were over the top—that was almost universally accepted by then—but whether that over-the-top attitude had crossed onto the news pages. Cape Wind had given Jurkowitz twelve news articles the company felt were problematic. Of those, the media critic dismissed half outright, believing them to be acceptable. Of the remaining six, he found three with what he considered "serious problems."

When his piece finally appeared in the magazine *Commonwealth*, Jurkowitz wrote that opposition to the project was "a view hammered home relentlessly on a take-no-prisoners editorial page determined to strangle the wind farm idea in its cradle," and a subhead asked "Has the *Times* launched a journalistic jihad?"

The critic wrote that the news pages and the editorial pages sometimes played off each other, ignoring the separation of powers between news pages and editorial pages that is considered essential to responsible journalism. "In boxing, that would be a three-punch combination," he wrote, adding that the editorials "have been insistent and more than occasional." Vis-à-vis the Patrick-Wheatley election, Jurkowitz confirmed that the newspaper's editorials were relentless. He gave plenty of print space to Schechtman and Meyer, who insisted their coverage had been fair on all accounts. "We've taken a strong editorial position on this," Meyer told the media critic. "But as strongly as we feel about the wind farm from an editorial position, we feel more strongly about . . . our editorial integrity."

The magazine piece appeared in April 2003. Jurkowitz later wrote two other pieces on *Times* coverage for the *Boston Globe.* He also discussed the issue several times on a local weekly public television show called "Beat the Press."

The *Cape Cod Times* did not take the criticism well.

"You're being used," Schechtman complained at one point to the veteran media critic.

He's just trying to embarrass me, trying to throw me off kilter and make me withdraw the story, thought Jurkowitz. Rather than being frightened, the remark further piqued his interest. He found that the paper continued to insert itself dramatically into the issue. "The paper's own institutional voice was relentless, constant, combative, to a degree which I would say was a little overwrought," Jurkowitz said. He was particularly troubled by the fact that Schechtman participated in writing some of the virulent editorials while also heading the newsroom. This created a "blurring of the lines between the paper's editorial views and its news coverage. Cliff Schechtman did not want to be a country editor. He had a view of himself on the paper as an activist. He was a very ambitious guy who wanted to exercise his power as he saw fit," Jurkowitz wrote.

When the media critic asked Schechtman why he had never sent a *Cape Cod Times* reporter overseas to look at an offshore project and to interview people who lived nearby, the editor said that whatever the reporter might find over there wouldn't be relevant to Cape Wind. Jurkowitz found that attitude indefensible.

"You can't have it both ways," he said. "You can't consider it the biggest issue in the neighborhood and not do what the smallest weeklies on the Cape did—send a reporter."

CHAPTER NINE

The Buzzards Bay Oil Spill

We get the oil spills, and they get the view.

John Bullard, Coalition for Buzzards Bay

The media critic's 2003 *Commonwealth* story created a buzz on Cape Cod, but it was quickly overshadowed by another event that also focused interest on the local energy controversy.

At the end of April, on a nasty, windblown night, a barge bringing oil to Cape Cod's power plant veered off course and struck a ledge, gashing its single hull. Almost 100,000 gallons of super thick toxic number 6 oil, also called Bunker C, found its way into ocean currents, eventually spreading a sheen over ninety miles of sandy shoreline, covering sea life, killing endangered animals like roseate terns, and making many beaches unusable for months.

The water smelled like a Texas oilfield. This could have been seen as the wrath of a righteous God, save for one problem: The beaches assaulted were not those of the southern-shoreline crowd. They were those of Buzzards Bay, a genuinely fragile, shallow, and sandy estuary

Oil covers the beaches of Buzzards Bay after being spilled from a barge.

west of Cape Cod. Buzzards Bay had already had more than its fair share of insults from oil.

However, west of Cape Cod sit two sizable cities with large numbers of people who live below the poverty line—New Bedford, on Buzzards Bay and known as "the City That Lit the World" because of its nineteenth-century whale-oil industry, and Fall River, on Mt. Hope Bay and known as "the Spindle City" because of textile mills that existed there long ago. Most of the residents of *these* cities are, horror of horrors, forced to summer where they winter. This group had a whole lot less political clout than the Oyster Harbors crowd.

The morning after the spill, Chris Reddy hitched up his low-ride jeans. He put on his old T-shirt filled with holes and, over that, a wrinkled shirt with a button-down collar. He wore psychedelic running shoes, just to make a point. Some people thought chemists were nerdy. No one would think that about Reddy. An ebullient and muscular man who enjoyed participating in sports like competitive wrestling and long-distance ocean kayaking, but who mostly lived for his science, marine chemistry, Reddy was high with anticipation. He could barely sit still.

It wasn't as if he *wanted* this spill to happen, but the prospect of data points flashed before his eyes the way prospects of big-money trades light up the eyes of futures traders. Reddy, a thirty-four-year-old researcher at the Woods Hole Oceanographic Institution with a slew of academic awards, was an oil-spill fanatic. Sometimes when he couldn't sleep at night, he thought about all the energy tied up in the chemical bonds that bound atoms together into the various oil compounds. He was a journalist's delight. When he talked about oil, he couldn't help himself. Any question asked would get a mile-long answer.

"Number Six oil is not like a can of Coke," he said, waving his hands in his laboratory office. "It's not a formula. It's the dregs left over in the refinery, after everything else is boiled off. The carbon that's left in there, the residual, down at the bottom, still has energy in it, so they sell it to power plants. The key thing to remember is that you have the dregs of stuff that you could burn—but it's a little too viscous, too thick. Often they cut it with something that's a little more volatile, something to make it a little more flowy. Refineries are like good old swamp Yankees. Whatever they have lying around that the market doesn't want, that's what they'll use to make the residual flow."

"They cut it! They cut it!" His hand did a karate chop in the air. Talking analytical chemistry with Reddy is not unlike talking about the Super Bowl with a football fanatic.

Reddy had cut his scientific teeth on one of the largest New England oil spills ever: the January 1996 North Cape disaster, a mammoth spill that spread 828,000 gallons of oil just to the west of Buzzards Bay in Block Island Sound. That spill closed 250 square miles to fishing.

Reddy was then working on his doctorate under James Quinn of the University of Rhode Island. His topic was a study of the fate of compounds in the ocean environment—then a scientifically glamorous topic, because it was red hot with controversy. After the 1989 *Exxon Valdez* disaster spilled 10.8 million gallons of oil in Alaskan waters—the largest oil spill in U.S. history—oil-company lawyers claimed the toxic oil would disappear with no long-term effects. They based the claim on laboratory tests that showed that the chemical bonds holding together oil's complex rings and chains of carbon gradually fragmented and became harmless.

Some environmental scientists, however, demanded corroborating field research. What happens in the laboratory isn't necessarily what happens in an ecosystem, insisted the dissenters. With the North Cape disaster, Reddy saw his opportunity to acquire some samples of the spilled oil and do his own study.

An endangered piping plover struggles for survival after being coated with toxic oil during the Buzzards Bay oil spill.

In one sense, oil is generic: long strings or circles of carbon are joined together, and compounded with other atoms attached on the sides or on the ends of the carbon chains. But a close analysis shows that oil has fingerprints. Each batch of oil has its own individual subsets of compounds. By identifying those subsets, a skilled researcher can pinpoint an individual oil. No matter how much a refinery tries to make a uniform product, each batch is unique.

One large group of compounds in oil, a classification called the PAHs (polycyclic aromatic hydrocarbons), is particularly toxic. These compounds harm the nervous system. In large enough doses, they kill. Lower doses cause convulsions, rapid heart rate, headaches, tremors, and a whole other list of life-threatening symptoms. The more PAHs in an oil, the more dangerous it is. Although some government-sponsored analysis of the spilled North Cape oil said it contained 3 percent PAH, Reddy's independent research put the figure at 7 percent. At first, no one believed the graduate student's findings. But Rhode Island environmental officer Stephen Morin followed up and found that Reddy was correct. It turned out that the barge contained two distinct batches of oil, one of which was substantially more toxic than expected. Reddy's science helped the state increase the settlement paid by the barge company. Morin praised the young researcher profusely. Reddy became something of a hero for not backing down.

In the fall of 2002, the limelight shone once again on Reddy when he found that traces of oil spilled in another disaster more than thirty years earlier remained nearly intact—and almost as toxic—under a thin layer of mud in a salt marsh. On September 16, 1969, a barge carrying highly toxic diesel fuel veered off course, struck rocks, and leaked 189,000 gallons into Buzzards Bay. Diesel fuel has lots of PAHs. The stench of oil was everywhere. For weeks, each tide brought windrows of dead sea life—lobster, flounder, scup, bay scallops—to Cape Cod's western shoreline. Locals called it the "silent fall." The bay looked like a biological desert, said one scientist. Sea life "declined from about two hundred thousand animals per square metre to about two animals per square metre," wrote *New Yorker* writer William Wertenbaker.

Woods Hole scientist Max Blumer, documenting the destruction, showed that in real-life situations oil behaved differently than it did in the laboratory experiments. Although the lab experiments showed that the bonds in oil could break apart quite quickly, making it biologically harmless, Blumer showed that in real life, what scientists call "in situ," oil could be reclusive, sometimes hiding in nooks and crannies, lying in wait like the genie in the lamp for an opportunity to escape. For years, when people stepped in marshes where the oil had hidden, in their water-filled footprints appeared a telltale rainbow sheen, evidence of the oil below.

The data amassed by Blumer and others over the decades provided the science-based foundation for regulating ocean transport of oil. Beginning in 2000, Reddy followed up on Blumer's science by researching Blumer's old sites, in a sense becoming the senior researcher's scientific grandson. No one had looked for the oil in more than a decade when Reddy's student Aubrey Hounshell began digging. It didn't take him long to find it. Analyzing Hounshell's samples, Reddy found that "the oil looked almost like fresh diesel fuel. That's the big question," he said. "Why hasn't this oil degraded more?" The toxins were there, tucked away under a thin layer of marsh mud, still affecting an animal that might try to burrow or peck its way down into the mud.

He talked about the life cycle of oil. "Plants die and get deposited in the sediments. If they get deposited in the right conditions, that old debris gets cooked and squeezed and made into oil." By the right conditions, Reddy meant that geological events like tectonic plate shifting sometimes force the layers of plant debris deep into the earth, down to a region roughly 6,000 to 15,000 feet below, known as the "oil window." There, heat and pressure break the plant compounds' complex bonds. The oil then sits for eons.

Reddy said: "Then it gets tapped and brought into a refinery. That oil represents many different types of compounds."

Some are light enough to evaporate. Other oil compounds are heavy, creating the beach tar that darkens toes for days after.

"The refiners catch that gasoline, kerosene, and jet fuel, the oil you put in the car. It's a sliding scale of viscosity. When they're done heating all that stuff and all the good stuff is evaporated away, made into gasoline and other products, you have the residue in the bottom of the pot. The oil that spilled was number six."

"Number Six—its composition can vary dramatically. Its cutting agent is anything that the refiners had that would make that residue easy to handle . . . ." Number Six oil, Reddy explained, can be fairly benign. Or it can be very nasty.

The morning following the spill, Reddy and chemist Bob Nelson, a large man with a persistently pleasant nature (even when he's been up all night analyzing samples), saw the opportunity to lay the foundations for another long-term study. They planned to cruise Buzzards Bay, taking samples from the large rafts of oil bobbing in the saltwater. They hoped to keep a detailed record as the oil spread in order to understand the oil's pattern of decomposition—its "fate"—over time.

For Reddy, what made this study particularly exciting was the presence of the GCxGC in their lab. The tool's official name is the Comprehensive Two-Dimensional Gas Chromatograph. When Reddy first saw the GCxGC in another scientist's lab, he was more than envious. He was in love. He felt as though he'd seen a very fast car, one that he absolutely *had* to drive. "Holy shit!" he said to everyone for days after first encountering this vision of loveliness. "You know what we can do now? We can tell a story about every compound that's in the oil!" This made his fingers dance on the conference table. "I've always wanted to do this—and now I can."

A chromatograph is to an analytical chemist what a hammer is to a carpenter or what an instrument is to a musician. Without a chromatograph, a chemist can't analyze, can't sort and separate compounds. The past century of progress in analytical chemistry is, to a very great extent, the history of the refinement of the chromatograph. With each small step in improvement in chromatograph technology, analytical chemistry has taken a giant step for humanity. The gas chromatograph used by Max Blumer, then state of the art, could not sort out oil compounds with any kind of precision. Reddy's supertool—the GCxGC—provided a hitherto unimaginable clarity. "Now we can tell a story about every single compound that's in the oil," he said. "In the past, we could only tell a story about 10 percent of the oil."

Buzzards Bay estuary was, at one time, before all the oil spills, an important nursery for the ocean. Its 228 square miles thriving with life astounded Bartholomew Gosnold when he sailed into its waters in 1602. Until 1914, the twenty-eight-mile-long estuary, averaging only about eight miles in width, was closed at its upper, eastern reaches. Ships from the rich cities of Newport or Providence sailing to Boston had to sail around Cape Cod, through the wide open Atlantic, often passing through Nantucket Sound. But in 1914 the Cape Cod Canal opened and ships changed course, taking a shortcut through the delicate and dangerous Buzzards Bay estuary.

Today, Buzzards Bay is the main shipping channel for about 600 barges carrying more than 2 billion gallons of petroleum products annually. "It totally took my breath away when I heard that number," said Reddy. Were it not for the canal, some of that transported oil might have passed through Nantucket Sound, but now the people of Buzzards Bay bear the burden of oil transport, leaving Nantucket Sound its illusion of pristine timelessness.

The estuary is a dangerous place for ships and boats when the wind comes up. The waves might not be high, as in the North Atlantic, but they're plenty steep and they just keep coming and coming. "The bay is pretty notorious for its chop," said Bob Nelson. "There's all that energy coming in and getting funneled into a smaller and smaller area. When the wind blows, the waves have a short period and a steep face."

When the scientists went out on their boat around noon, they thought they were in for a decent afternoon. "Not a bad day for a spill," Reddy joked. The water seemed smooth. The sun seemed warm enough, if not particularly hot. But by midafternoon, the wind was up. The water roughened and the boat took it hard. The scientists were

Scientist Chris Reddy examines the health of marine organisms during the Buzzards Bay oil spill crisis.

hammering through the chop. At each sample point, Nelson had to lean out over the gunnels to grab a handful of glop out of the rafts of oil floating in the swirling sea. "At every station," he said, "it got rougher and rougher. We had to have someone either holding my belt or grabbing my legs to be sure I didn't fall in."

Nelson wore surgical gloves to keep his chemical fingerprints off the sample. "You don't want to be contaminating the oil," he said, and then upon reflection added, "and you don't want the oil contaminating you, either."

Still, the oil on top of the waves kept splashing up in his face. "We were in some pretty intense fumes. There was a distinct smell of oil out there, like asphalt. The next day I had a bloody nose." By midafternoon, they both had headaches.

Around 7 P.M., when darkness forced them to come in, they were overwhelmed from having breathed in so many toxic fumes. After taking some time to answer the questions of reporters waiting at the dock, the team headed back to the lab for a night's work. After a couple of hours of prep time, separating the oil from the other junk grabbed up from the oil rafts, they began processing these first samples. By midnight the first data appeared on the computer screen.

The results surprised Reddy and Nelson. The oil contained a high percentage of naphthalenes—the insecticide in mothballs, and a subgroup of the toxic PAHs. You don't want to inhale too much, because in large enough quantities, the same poisons that kill moths can kill you, too.

What happens to those naphthalenes in oil after a spill, Reddy wondered? Where do they go? How long do they remain toxic? No one knows the full answer to that question. Some are probably eaten by a common little fellow, *Pseudomonas putida*, a bacterium that makes its living by eating hydrocarbons that occur naturally in the ocean, he said. Reddy saw an oil spill as an all-you-can-eat banquet for bacteria. "Oil is like butter," said Reddy. "Some bacteria love to eat it." On the other hand, the bacteria don't clean up the whole mess. Some of the naphthalenes evaporate. The energy from sunlight sometimes breaks apart the chemical bonds, reducing toxicity. "When they're up in the air, a lot get whacked by photons," said Reddy. "Then again, we don't really know . . . "

Before sunlight or some other force breaks up the toxins, the compounds could just as easily be wafting over the shoreline, into someone's lungs. An odiferous, unpleasantly yellowish haze hung over Buzzards Bay the day after the April 2003 spill. Lots of people reported headaches, not just Reddy and Nelson.

Reddy found about sixty different types of naphthalenes, not an unusual figure. But there seemed to be a greater percentage of the toxic compounds in the oil than he'd expected. Given this high percentage, he predicted that an unusually high amount of sea life would die. He turned out to be correct. Only 10 percent of the oil-covered birds taken to cleaning centers survived.

Birds suffered in other ways, too. Buzzards Bay estuary is the site of North America's largest colony of endangered roseate terns. Since the spill occurred at nesting time, volunteers had to go to the estuary's islands where the animals were mating to scare the birds away. To frighten the birds they blasted sound-cannons that could be heard as far away as the mainland.

Shellfish beds were ruined. Because the estuary had been better protected than Nantucket Sound from certain destructive fishing techniques, the shellfish beds there were in good condition. When Bobby Kennedy Jr. claimed that Nantucket Sound was the region's "economic engine," he may not have been conversant with the ecology of Buzzards Bay, which nursed a $4 million shellfishery, about 25 percent of the state's total. The estuary's communities annually issue more than 500 commercial permits and nearly 13,000 recreational permits. The Oyster

Harbors crowd complained that tourists would avoid Nantucket Sound if the wind farm were built, but after the oil spill, tourists wanting to dig up their own clams for chowder were out of luck on Buzzards Bay.

That same evening, while Nelson and Reddy were running their samples, Jim Gordon settled onto his couch with his wife to take in the late-evening news. The first video images of the oil spill came as quite a jolt. Gordon thought about the impact the spill would have on the coming summer tourist season. He also thought, somewhat selfishly, he admitted to himself, that this was exactly the kind of choice he had been presenting—dirty power based on continued use of fossil fuel, or power produced from the wind, using twenty-first-century technology.

The tragic spill changed the nature of the Cape Wind discussion in a way no mere debate of words ever could. The context widened, so that Cape Wind was no longer seen as just a local issue. For several years, Cape Wind opponents had been saying the project should be put "somewhere else," but now, the narrow focus of many project opponents suddenly seemed myopic. Up until the spill, for many people, the debate had been theoretical. Smokestack emissions are invisible. Gordon sometimes yearned for a forensic spray that could make the small airborne pollutants visible, to help people actually see the effects of burning fossil fuels. But everyone could see this spill's black, gloppy goo, which clung to everything like a disgusting version of saltwater taffy.

One of the worst-hit areas in the bay was West Island. Residents there are accustomed to walking their shorelines and looking at the almost continuous line of ship traffic heading over to the Cape Cod Canal—but they aren't used to oil barges sitting still. After the spill, they looked out directly at the damaged oil barge. It made people uneasy. A stream of helicopters flew overhead, cruising so low that the windows of the island's cottages rattled constantly.

A member of the oil spill cleanup crew uses high-pressure steam to clean up spilled oil on West Island.

As the helicopters monitored the spread of the oil, residents walking outside to watch smelled a vague aroma wafting over the water. Media reports of the oil spill were by now coming over the radio, but the reports confused islanders. First they heard that the barge had spilled only a small amount of oil, nothing to be alarmed about. But as the hours passed, the amount spilled kept rising, from about 9,000 gallons to about 15,000 gallons.

The next morning the situation remained unclear. It was a beautiful day, the rare kind of special early-spring morning that feels magical to New England coastal dwellers who have weathered the depressing gray winters brought by the North Atlantic. That day, the sky was a promising gentle blue. But the ocean looked ominous. The water was a metallic gray-and-black with a gel-like quality. As the oil drifted in, the smell was overpowering.

Across the island, residents began getting headaches and closing their windows, trying to escape the fumes. The ironies for the people of

West Island, many of whom had paid little attention to the Cape Wind controversy, were profound. When the connection between the Nantucket Sound battle and the Buzzards Bay oil spill became apparent, many people were angry.

The historic parallels between Oyster Harbors and West Island are uncanny. Both islands are small, requiring only a few hours to walk around at low tide. Before European settlement, native peoples summered on both islands, taking refuge from summertime insects and heat and enjoying the access to good shellfish beds. Europeans turned both islands into farming and grazing land. Then real-estate developers converted both islands into summer-housing developments. Both islands are low, only a few feet above sea level, vulnerable to climate change, and would be unsuitable for year-round habitation were it not for electricity, cars, and telephones.

There is one key difference, though. While Oyster Harbors is building a new 44,000-square-foot clubhouse for its members, the community on West Island has made do for decades with a small bunkerlike meeting house made of concrete that can barely fit 150 people. While Oyster Harbors has cocktail parties on a large verandah, West Island has a once-a-month bacon-and-pancakes community breakfast. People sit at picnic tables. Oyster Harbors residents call their view of the sound "pristine." West Island residents know very well that their water is far from pristine.

While the Mellon family worked hard to make Oyster Harbors exclusive, West Island during the 1950s was available to anyone brave enough to buy a $200 or $300 lot. When Terry Pereira of New Bedford, one of Massachusetts's poorest cities, married in 1957, her father-in-law bought the young couple a West Island lot. A city girl, Pereira hesitated about living on West Island, but her husband, a first-generation New Bedford–born Portuguese American, wanted his children to grow up near the sea. At first the young wife felt like a pioneer, particularly when her husband and father-in-law built their family house themselves, with their own hands and their own tools. Always treasure the land, her immigrant father-in-law instructed before he died.

The *Bouchard* spill horrified Pereira. Like others on the island, she knew nothing about the spill until she smelled the oil. Thinking it must be coming from a nearby harbor, she went down to the water's edge to look and saw birds covered with oil, then the slick itself. She realized immediately that her ten grandchildren, ranging from toddler to early teens, wouldn't be able to spend this summer on the water. The few times they tried, they came back with oil all over their shoes or their bare feet. The Coast Guard handed out baby oil to help get rid of the tar on the kids' feet, but it didn't really work. The poisons stayed on the children's feet.

Like other islanders, Pereira learned that the spilled oil was intended to fuel Cape Cod's electrical plant, but she had not heard about Cape Wind and had no idea that the proposed wind farm might affect her in any way. When someone told her that the field of wind turbines suggested to be built far to the east of her island would likely decrease the number of oil barges passing her home but that Cape Codders were fighting the project, she just shook her head. To have her now-beloved island defiled and to have people on Nantucket Sound refuse to allow the wind project to be built was more than she could understand.

Also frustrated and angry was a tall, thin, red-haired young woman who had only a half year earlier given up a high-paying job as a patent attorney to practice law for a small advocacy organization called the Coalition for Buzzards Bay. Susan Reid, thirty-two, was at a meeting in San Francisco when the spill occurred. On her way home, a friend called to tell her what had happened. When Reid finally saw the oil blanketing the beaches where she'd played as a child, she felt as though she'd been shot.

No one seemed to be able to stop the oil from covering the marshes, smothering the small organisms that lived there, poisoning the birds. Pools of oil gathered everywhere, in the sand, the rocks, the marshes.

Seeing the spill, Reid felt an overpowering sense of suffocation. She felt helpless to protect this childhood favorite place, but she also felt hypocritical. After all, she herself drove a fossil-fuel car. The crisis gave her a new sense of determination.

The spill-recovery teams were unprepared for the amount of oil covering the beaches, which had turned out to be at least 98,000 gallons, far more than the barge company had first estimated. To Reid, their efforts seemed so ineffectual as to be almost useless. Reid saw recovery-team employees, dressed in orange suits that made them look like men from space, kneel down amid the puddles of oil and try, with tongue depressors, to remove the toxic glop from acres and acres of delicate marsh plants. How could they accomplish such a Herculean task, she wondered, with a tool like a tongue depressor?

On the other hand, some orange-suited employees pointed high-powered hot-water hoses at acres upon acres of oil-covered rocks, trying to wash the poison off. This large-scale solution frightened her, since there seemed to be such a potential for so much ecosystem damage. In this case, at least, it seemed as though the cure might be worse than the affliction.

Reid pushed government officials hard, trying to force them to do a better job. When they declared an area clean, Reid knelt in the sand and dug down herself, finding the still-present oil below the surface. Officials and barge company representatives came to hate her, which bothered her, but she kept on pushing. She felt all the parties—*Bouchard* officials, government agents, cleanup companies—were trying to just wash away the visible oil and skip out. Reid was offended that the barge company could behave recklessly, whitewash the problem, and spend the least amount of money possible. If they thought no one would notice, Reid would make sure that wasn't the case.

The disaster profoundly affected Reid's attitude toward Cape Wind. Before the spill, the young attorney had listened with half an ear to the arguments on Cape Cod. The sheer size of Cape Wind's proposal seemed intimidating. Her organization, the Coalition for Buzzards Bay, had tried to keep at arm's length from the debate, in part because it belonged to the Waterkeeper Alliance, a coalition of environmental groups advocating for clean water—headed by Robert F. Kennedy Jr. Kennedy had tried to force the Waterkeeper Alliance to oppose Cape Wind. He failed, but did succeed in sowing seeds of doubt.

On April 30, three days after the spill, when the mess on the estuary's beaches was worse than ever, a *New Bedford Standard-Times* editorial asked: "Where are Romney, Reilly in hour of need?" The editorial fervently denounced the state's governor and attorney general, both of whom had shown little interest in the spill but had persistently and loudly decried Cape Wind's "environmental devastation."

"As fuel continues to lap up on our shores, we are angered by the fainthearted response of our state's top leaders," the editorial said. Both men had often claimed Cape Wind would "desecrate" Nantucket Sound, the editorial reminded readers. "But this spill is real, immediate desecration of Buzzards Bay." Where were the politicians? "Why aren't they on the beaches of Dartmouth and Fairhaven?" Good question, thought Reid.

Eventually, Romney made an appearance—a carefully crafted made-for-television appearance on Barney's Joy beach, a private beach with expensive membership fees, where the governor would not have to talk with any but the wealthy folk who maintained the area. Reid wondered about Romney's motivation. She knew that the governor had been asked to visit the out-of-work shellfishermen of Hoppy's Landing, near West Island. Romney had refused. Reid wondered if he'd not wanted to appear in locations where he might encounter upset working people who might ask difficult questions.

In 2002, the canal electric plant burned about 7 million barrels of number 6 oil. That figure translates into somewhere between fifty-eight and seventy oil barges passing through Buzzards Bay each year, so that Cape Cod—including its many energy-greedy huge shoreline homes—could have electricity. Also in 2002, Massachusetts's longtime energy czar David O'Connor wrote that "the Cape Wind units would be more likely to displace the operation of the Canal plant than any other plant, but when and by how much is impossible to say."

Gordon had not done a good job of getting that message across, but the spill did the job for him. In addition, supporters began to look more extensively at the project in the very wide context of the full power grid. New England's grid spreads from Cape Cod to the New York border, from northern Maine to the Connecticut shoreline. What happens in one area of the grid affects all New Englanders.

John Bullard, the board president of the Coalition for Buzzards Bay and a former New Bedford mayor—who traced his ancestry back to the earliest days of whaling when Rotch family members came over from whaling-rich Nantucket in 1765 to establish a deep-water mainland port—was among those who began to speak out. Bullard and his wife still live in the 1845 house in the center of New Bedford built by their whaling ancestors. They remain loyal to their city, despite its current difficult times.

A dedicated Democrat, Bullard was profoundly disappointed with Ted Kennedy, whom, he said on a New England–wide television show, he would no longer support because of Kennedy's selfish stance in opposing Cape Wind. He was also disappointed with Romney, whom he accused of lacking political courage and genuine leadership ability. Buzzards Bay, Bullard was convinced, was as much a "national treasure" as Nantucket Sound.

"Romney and a lot of other politicians think supporting the wind farm is a losing issue politically," he said after the spill. "While I disagree with that, if you are looking at an issue you think is a political loser, then NIMTOF kicks in. You say, 'I'm for wind power, just not here.' What you really mean is, 'I'm for wind power, just not now.' If you translate political speak into English, into the truth, what they really mean is 'not in my term of office.' Most political leaders only want to evaluate the wind project in comparison to a pristine environment. That, of course, is not the case. There is smoke coming out of stacks today. There is oil in the water and in the mud today because of fossil fuel as an energy source."

"There's always been a justice aspect to this debate, and you can see that most clearly with the oil spills in Buzzards Bay, in Fairhaven and New Bedford, versus no oil washing up on the beaches of Nantucket and Martha's Vineyard and other parts of the Cape. You won't find PCBs in Nantucket Harbor, because industry has always chosen places like New Bedford to discharge their waste, while captains of industry have always vacationed in places like Nantucket."

"There are a couple of ironies here. One is that the rich and powerful are mistaken in thinking they can get away. Because there is no 'away.' The foulest air can be found on Mount Desert Island in Acadia National Park, and it's not because of the heavy industry on Mount Desert Island. It's because of the mills in the Midwest. The captains of industry have fouled their nests so thoroughly that even they can no longer escape."

CHAPTER TEN

The Senator from Virginia

There has never been a democracy yet that did not commit suicide.

John Adams

If the oil spill brought the Cape Wind controversy to the attention of people outside of Cape Cod, it did not cause project opponents to reassess their stand. Alliance leaders did not address the connection between the oil spill and the canal electric plant or between the power plant and the wind-energy project they opposed. Throughout 2003, money continued to flow into Alliance coffers. Alliance spokesmen called for volunteers to help collect injured birds and to help clean up the mess from the spill, then continued its opposition efforts on its many different fronts.

In early May, only days after the oil spill, local columnist Francis Broadhurst exposed what might have been the pettiest of project opponents' tactics to date. Months earlier, attorney Allen Larson, a Cape Cod Community College trustee and leader of a local environmental sustainability organization, had asked Gordon for a donation for the college's sustainability program. Larson had briefly represented Gordon during an early phase of project development.

Gordon agreed to donate $100,000 to help fund a renewable-energy curriculum. College president Kathleen Schatzberg had announced the donation at the February 2003 meeting of the college trustees, only to encounter strenuous objections from two of the wealthier board members. Meeting minutes say that George Zografos, aka "the Doughnut King" because of the Dunkin' Donuts franchises he owned, and Paul Swartz, owner of a hotel on Yarmouth's honky-tonk strip, claimed that accepting Gordon's no-strings-attached donation would be tantamount to supporting Cape Wind.

Other trustees, including, of course, Larson, disagreed with that point of view. Larson thought the donation could help the college assume leadership in emerging energy technology, thus gaining national recognition. It seemed to him a no-brainer. The college could be way ahead of everyone else as an institution, if it so chose. Students could be offered career opportunities and technical knowledge that could help them earn improved wages, and it would be in sync with the Cape's stated goal of sustainability. Schatzberg decided to turn Gordon down.

Francis Broadhurst occupied an unusual position as a regular columnist for the *Cape Cod Times.* Decades earlier, he had been a selectman in Barnstable, and when he left politics, he had been hired by the paper's preceding editor, who felt that Broadhurst's very conservative columns represented an important segment of Cape Cod society that was underrepresented in print. The aging writer had a large and very loyal following of longtime Cape Codders, many of whom were born on Cape Cod, or who had arrived long before the most recent wave of Cape Cod's new-money folks. From the outset, Broadhurst supported Cape Wind. *Times* editors had refused to run some of what he wrote, but he was too popular with readers to be fired outright.

When his column reported that the college was refusing to accept a donation, it started a community firestorm. Under Romney, state college funding had been slashed. Staff had been dismissed, professors had heavier teaching and other work loads, and even subscriptions to magazines like *Science* and *Nature* had been stopped. A pro-Wind group of local organizations published a statement: "We believe the fractious debate over renewable energy development in the Cape & Island regions has just crossed an important threshold: Possible misperceptions by the public about the Cape Wind project are being used to sway individual and community decision-making about larger issues, to the detriment of local interests."

The increased public pressure created an environment in which the college decided to accept the money after all, but the "fractious debate" was to continue on other fronts, large and small. In early summer, Cape Codders opened their newspaper to find a rather luridly colored map inserted as a paid advertisement. The caption in the map's center, in black-and-rust-red lettering, said: "Imagine waking up one morning and finding that a big chunk of Nantucket Sound had been stolen."

The map showed Nantucket Sound itself, surrounded by the devil's triangle of Nantucket, Martha's Vineyard, and Oyster Harbors. It contained an Alliance-drawn outline of Cape Wind in Nantucket Sound, locating the project several miles closer to the Cape Cod shoreline than it actually was. Readers would assume that the turbines would loom vulturelike over Cape Cod's beaches, whereas in truth, the turbines would be located roughly six miles from the Cape's southern shoreline. Moreover, the project's footprint was substantially enlarged. Cape Wind claimed the map tripled the size of the actual proposal.

The main thrust of Alliance opposition, however, seemed to be moving gradually to Washington. Help came from an odd quarter—U.S. Senator John Warner, a Republican from Virginia.

Warner had become involved in the seaside civil war as early as the summer of 2002, while Jim Gordon was in the midst of getting permits for a seemingly innocuous installation in Nantucket Sound of a platform and monopole bearing scientific instruments. The instruments were to gather data on wind speed that would show whether his proposed field of turbines would generate enough electricity to make the project economically viable.

During his Wianno Club presentation, Alliance head Doug Yearley had promised to target the permitting of this tower. Opponents chose to do so, because without this information, Cape Wind would not be able to go to the financial markets to raise capital.

Project opponents saw the data tower as much more than just a simple skirmish in the wind-farm wars. In their view, this tower presented an opportunity for a preemptive strike, for shock and awe, an opportunity to let Gordon know that they meant business in opposing his project, and that they were willing to put up the money to back up their opposition.

When planning any wind project, on land or on sea, developers need a vast amount of very precise data that provides hour-to-hour information on wind speed at various heights, wind gusts, directional shifts, air turbulence, and much more. Planning an offshore project also requires that developers know about water temperature and air temperature, about undersea currents that might erode turbine foundations, about wave height, wave period, and wave direction.

A small amount of that data already existed. Only a few miles away from Horseshoe Shoal was Nantucket Island's small airport, used mostly by private planes, which had ten years of wind and weather records. Cape Wind had purchased those records, which it hoped to correlate with its own data collection.

Should the acquired data be favorable to moving the project forward, it would be part of the documents used to raise money in the financial markets, by providing a fact-based prediction of how much electricity the project would produce. Because it was to be used in the financial markets, accuracy was essential. The power produced by a wind turbine is the product of the wind speed—*cubed*. This meant that an error of only a mile or two an hour in the average wind speed would mean a substantially altered estimate of power output. Data needed to be as complete as possible. Capital markets would not buy into vague estimates or haphazardly acquired information.

Permitting a data tower was a routine endeavor overseen by the Army Corps, so Gordon was expecting smooth sailing when the company filed its notice of intent with the federal agency in December 2001. He promised to make some of the gathered data, which would cost around $2 million to acquire, available to the public.

Gordon hoped to have the tower in the water by early summer 2002, but the Corps lengthened the permitting procedure by opting for an unusual addition to this normally routine permitting. Having received calls of "concern" from staff of a long list of politicians, including Mitt

Boats encircle Cape Wind's data tower during an anti–Cape Wind protest on Nantucket Sound.

Romney and Ted Kennedy, the Corps decided to hold a public hearing on the Cape.

Under the circumstances, Corps staff wanted a clear record of having encouraged community involvement, in the all-too-likely event that the issue ended up in court. The April 2002 meeting at Barnstable Town Hall was attended by all the usual suspects, who made all the usual charges about whales, seals, and so on. "I question why this project is still alive!" a red-faced Barnstable councilor Gary Blazis sputtered, to no one's surprise. Someone complained that the data tower was "the camel's nose under the tent." Barnstable councilor Bob Jones insisted that there were "hundreds of issues" that made the full-scale project unviable, and cited a laundry list of "environmental harms."

Gordon sat on a folding chair in the front row right next to the microphone. Dressed in a work shirt instead of his usual suit and tie, he had adopted the folksy outdoor-man dress of Cape Codders. He kept his arms crossed and listened carefully. Behaving something like an inanimate object, for most of the evening his facial muscles barely moved.

Project supporters had become more organized by this meeting, and there were many speeches favoring Gordon's proposal, but the most striking was the performance of an older man whom nobody knew. "I live in O'ville," said Russ Haydon, adding that he was sick and tired of hearing all the "hysteria" and all the "baloney" coming from the anti–wind farm folks. "Wait until you get the facts," Haydon said.

Then he turned away from the microphone and pointed toward the audience, accusing them of behaving like Joseph Goebbels, head of Hitler's propaganda machine. "He knew that if you told a lie frequently enough, often enough, people would accept that, believe it! You don't want facts! You don't want information! Unfold your arms! Chill out! We've had enough propaganda!"

To Gordon's immense frustration, the Corps followed the hearing with several months of consultation with other agencies regarding potential environmental impacts. By this time, the tower installation had been delayed by six months. Gordon figured that, even with that delay, by hurrying he could get the structure in the water before the rough winter weather set in.

He hadn't, however, counted on the services of Senator John Warner, senior Republican on the Senate Committee on Armed Services. When the Corps was about to issue the permit, Warner made his public debut in the wind-farm debate by demanding yet another delay. In an August 1, 2002 letter written on Senate stationery, Warner asked Les Brownlee, then head of the Corps, to "withhold a decision" on the tower permit "until such time as I and other interested colleagues can be briefed by you on the implications of the entire project."

Under his signature, Warner included a handwritten, personalized message: "P.S. I have visited this area for over a half century—it's a national treasure." Reading the "national treasure" mantra, Gordon and his supporters rolled their eyes. How could this Alliance slogan have turned up in a handwritten note from a Virginia senator? Congress funds the Army Corps of Engineers. If a U.S. senator intervenes in a project, the Corps pays attention, particularly if that senator is the head of the Senate's Committee on Armed Services. Project supporters braced themselves for a long delay, but in fact Brownlee scheduled the meeting quite quickly.

Following his meeting with Brownlee, Warner wrote a second letter, dated August 16, filled with the usual hyperbole. "The proposed wind turbine project is of colossal proportions for the fragile ecosystem" of Nantucket Sound, wrote the Virginian senator. "Colossal," of course, is a relative word. To some the project was colossal, while to others it was not large enough. "Fragile" was just silly. "Fragile" means delicate and easily damaged. Roiled constantly by ocean currents and severe storms, Horseshoe Shoal was anything but fragile.

Warner also wrote that Cape Wind had no right to build the project in national waters because the company did not own or have property rights to those waters. He closed with the usual threat of legal action.

"Interested parties will have to decide if there is a basis to now seek injunctive relief in the federal courts to further delay the project," he wrote, following up on Yearley's promise of endless litigation.

Warner also made clear that he intended to stay involved in the seaside civil war: "I will continue to follow this matter."

And indeed he did.

Warner's obligations traced back to Bunny Mellon, his former mother-in-law and longtime financier. Explained Francis Broadhurst: "When my son worked for Fred Conant, who tended the gardens for the many Mellon homes in Oyster Harbors, they still called the one the senator stayed in 'the Warner House' even though he had since married Liz Taylor."

John Warner was thirty when he married Catherine Mellon in 1957. Author Kitty Kelley wrote that Warner "bragged" to a friend that his income from Mellon trusts was "$9,000 a day" as a result of the marriage. An advance man for Nixon during the 1960 presidential election, Warner subsequently held jobs in various Republican administrations, rising eventually to become Secretary of the Navy after the second Nixon election, a position that Kelley implied was due to a contribution of "$100,000 of Mellon money" to the 1968 election campaign.

What Warner really sought was elected office, a desire that was very much at odds with that of his wife. Catherine Mellon Warner wanted nothing to do with election campaigns or the duties of the wife of an elected official. The couple were also split over the Vietnam conflict. Warner wanted the United States to stay in the Southeast Asian country to "finish" the job. His wife wanted the troops home. "My wife was almost a student radical," Kelley reported Warner saying.

The 1973 divorce seems, by all accounts, to have been friendly. Warner is said to have received an $8 million divorce settlement from his wife's father, Paul Mellon. Warner also got the Georgetown house, custody of the couple's three children, and the 2,700 acre horse-and-cattle farm in Middleburg—next to what had been the estate of John and Jackie Kennedy and quite near Bunny and Paul Mellon's fox-hunting estate, where they had entertained Queen Elizabeth and her husband. He also received a substantial lifetime income, courtesy of Paul and Bunny Mellon. His former wife bought a house next to Warner's Georgetown house, and the couple shared childcare. Kelley reported Warner saying that Paul Mellon was "very generous to me. Through his generosity I am now able, and will be throughout my life, to pursue a career of public service."

Several years later, Warner married Elizabeth Taylor and began to run for public office. He was anything but a shoo-in for the job of senator from Virginia, and in fact lost his primary race. However, after his primary opponent died, he ran on the Republican ticket. During the election, Taylor seemed happy to show up at stockcar races and ladies' teas. The marriage to Taylor, longtime buddy to Kennedy clan members, raised some amused eyebrows and eventually resulted in a *Saturday Night Live* skit, wherein John Belushi played the role of Liz Taylor as a happy little political housewife. According to author J. Randy Taraborrelli, Warner "used" Taylor in order to get elected. Warner won the election by roughly 4,000 votes. Both Warner and Taylor agreed that Taylor could have pulled in that number on the strength of her own fame. After the 1978 election, the marriage gradually fell apart. Warner paid little attention to his famous wife, and, for her part, being a happy little housewife wasn't really her thing.

Despite life's ups and downs, Warner and the Mellons have stayed very closely connected. The senator attends family functions at Oyster Harbors, and the Mellon family continues to support him in various ways, including campaign contributions.

By insisting on a meeting with the head of the Corps before the Environmental Impact Statement was released, Warner had managed to delay the data tower for only a few weeks, but his efforts put Jim Gordon at odds with nature's deadline of heavy winter seas. Installation immediately began. Ten days into the construction process, two separate groups filed legal appeals to the Corps' decision, as promised by Warner. Along with the Alliance, a second group filed as a Ten Taxpayer Group, led by wind-farm opponent and attorney John W. Spillane. The Ten Taxpayer Group included the Massachusetts Boating and Yacht Club Association, a group with many members who could easily afford the legal fees.

Filing an appeal does not require a halt to construction, so Gordon had his crew continue working. When an attorney for the yachting group obtained a stop-work order in Barnstable County Superior Court, Cape Wind immediately went to federal court, which declared that Barnstable's county court had no jurisdiction in federal waters. Lifted on October 8, the Barnstable court order had successfully delayed installation by another ten days.

Doug Yearley had promised that the Alliance would focus on stopping construction of the data tower, but Gordon and project supporters were startled by the extent of the power of Cape Cod's wealthy folk. Gordon was getting worried. The Alliance's various strategies had delayed the tower's installation by more than six months, so that the company had to run hard to complete installation before the winter weather set in.

They succeeded, but remarkably, even after the temporary tower was in place, the attorney for the yachting club pursued his suit, losing each step of the way. He vowed to take his case all the way to the U.S. Supreme Court, which is exactly what he did. His petition was denied, but it wasn't by any means clear that the goal had been to win the court suits. Opponents had said repeatedly that they merely intended to keep Gordon in court. Yearley's strategy of "targeting" the data tower cost Cape Wind roughly an extra quarter-million dollars in legal fees.

Meanwhile, during this same period, scientists from the Woods Hole Oceanographic Institution installed a very similar temporary data platform and tower in the same region. No one said a word. It began to seem as though Yearley and Gordon were playing a game of high-stakes poker, and the first man to blink would be the man who lost.

When John Warner warned in his letter to the head of the Army Corps that he intended to stay involved in the Cape Wind issue, he meant what he said. During 2003, a flurry of legislative language floated around the halls of Washington, its authorship often unclear. At one point, a Warner spokesman admitted publicly that the Virginia senator had talked with Senator Kennedy and Representative Delahunt about blocking Cape Wind, but exactly which of the 2003 proposals involved Warner directly remained uncertain, and Warner wasn't talking. Both Kennedy and Delahunt were trying desperately to find some way to legislate some kind of moratorium, either specifically related to Cape Wind, or to offshore wind in general.

For Cape Wind attorney Dennis Duffy, who spent his vacation working the phones from his Martha's Vineyard home while glued to the C-SPAN-televised energy hearings, the 2003 efforts were only the beginning of many of what he would come to call "legislative-alert fire drills." Before Cape Wind was proposed, Duffy had had limited experience in Washington's halls of Congress. He had to scramble to find out what was going on in those legislative backrooms, but as he learned the ropes, he was surprised, and then heartened, to learn that Cape Wind had more friends than he'd realized. Once he explained the project to House and Senate staffers, he often found they loved the very grandiosity of the idea that had so upset Cape Wind's opponents.

A crucial test of Cape Wind's support in Congress came in October 2004. A defense appropriation bill had made its way through both the

House and the Senate and was in a conference committee that consisted of both the House and Senate Committees on Armed Services. Since the bill allocated money for troops in Iraq, passing the bill was urgent.

At the last minute, just as it seemed the bill was headed out the door, Warner, then head of the Senate Committee on Armed Services, added a rider that would prohibit the Army Corps from spending any money on permitting any offshore wind projects. A legislative source, which Cape Wind declines to identify, sent the company a copy of Warner's proposal.

When the legislation surfaced, Cape Wind was in a tough spot. To delay the progress of such an important bill was unacceptable. On the other hand, when the public realized that Warner was dragging American troops into the Cape Wind war, the opprobrium might cause the powerful senator to back down. By now Duffy was no longer a Washington innocent. He knew how to marshal environmental organizations to call congressional offices. He knew the unions would back him. He knew where Cape Wind's friends were, and how to marshal those friends quickly in an emergency.

Cape Wind's Washington friends had grown in number, as more and more people had become aware of the proposal and the nature of the obstacles the proposal faced. Normally in Washington power-plant proposals are regarded as local skirmishes, and congressmen and senators adopt a hands-off attitude because they don't want others interfering with their districts and states. But by late 2004, the picture had changed. The traditional collegial deference was disappearing as Cape Wind made national and even international news. It was hard for America to be taken seriously at international gatherings on energy while the opposition to Cape Wind was covered so heavily in the press. Moreover, the American public was beginning to get upset by rising energy prices, by an increasing understanding of climate change, and by the rolling blackouts that were threatening so many different electric grids around the nation.

After receiving notice of Warner's legislation, Cape Wind marshaled its forces. Blast e-mails were sent to unions, to environmental groups, and to energy organizations. Phone calls flooded the offices of members of both the House and the Senate. Clean Power Now, Cape Cod's pro-wind organization, visited congressional offices. Cape Wind communications staff e-mailed copies of Warner's legislation to newspapers, many of whom immediately ran stories about the senator's attempt at an end run.

Word came back that Warner was meeting with stiff resistance from Senator Carl Levin, the ranking Democratic senator on the Armed Services committee and a longtime supporter of renewable energy and of ending America's dependence on foreign fossil fuels. Levin simply believed that Warner's last-minute strategy of stopping offshore wind via a bill intending to finance American soldiers in Iraq was the wrong way to make energy policy. The other powerful holdout was Duncan Hunter, the Republican chairman of the House's Armed Services committee. Representing California's fifty-second congressional district in the southern part of the state, Hunter had received many awards from renewable-energy advocates and, like Levin, wanted to see the industry move forward.

Given all the stress about energy, and all the publicity Cape Wind was getting, it was awkward for Congress to allow Warner to get his way. The story of Warner's conference committee moratorium proposal hit the nation's newspapers on October 7, 2004. Even the *Wall Street Journal* carried the story. News that Warner had withdrawn the legislation appeared on October 8, 2004.

The following day, an editorial in the *Daily Press* of Newport News, Virginia, asked: "At what point did the senator take an interest in that particular coastline? Could it be that certain residents of Nantucket, a Mecca of relaxation for the high, mighty and financially well lubricated, got the word out that the 130 spinning things offshore would, well, detract from the refined aesthetics of it all?"

The power the Alliance held on Cape Cod apparently failed to impress the rest of the country. Warner did not come out of the skirmish unscathed. When Virginians learned about his efforts, they were not pleased. A Warner spokesman was reduced to denying that the senator's action had anything to do with his family history.

Later, Warner would tell Dennis Duffy in a Senate hearing room that he "came within a millimeter of getting my statute of a year's moratorium through. I won't tell you what happened, but one individual was able to stop it in the other body. That's the way we do business here. I'm not complaining, I've done it myself."

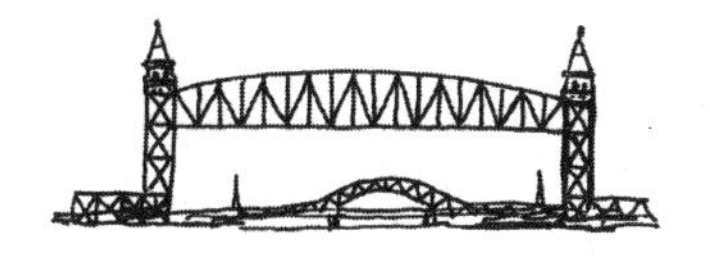

CHAPTER ELEVEN

A Cold Snap

Of all forms of tyranny, the least attractive and the most vulgar is the tyranny of mere wealth.

THEODORE ROOSEVELT

If the Alliance wasn't making any progress with its legislative forays, neither was Jim Gordon making much progress in getting his project permitted. The massive Environmental Impact Statement seemed mired in the data-gathering stage, and the Alliance and their politicians took advantage of delay after delay to wage their "valiant" battle for the hearts and minds of Massachusetts residents.

In early January 2004, Gordon was asked to participate in a radio debate with Ernie Corrigan, who had been bugging Boston radio host Howie Carr for months to do a Cape Wind show. Expecting more of the same annoying hyperbole he'd been listening to for several years, Gordon accepted, but not with great enthusiasm.

At the studio, Gordon learned that there would be a third participant—Theodore Roosevelt IV. Gordon was a taken aback. He hadn't known there *was* a living Theodore Roosevelt. Roosevelt, an investment banker from New York, had a summer home on Chappaquiddick.

Great, thought Gordon, feeling outgunned. He assumed the great-grandson of President Theodore Roosevelt would oppose the project.

By on-air time, Corrigan hadn't shown up. Carr began the show by explaining that the two fossil-fuel plants that provide most of the Cape's electricity belonged to what was known as Massachusetts's "Filthy Five." The wind farm, Carr said, would provide clean energy. But, Carr added with a little snigger, there's another side to the story.

"A lot of people who own property down there, they say that we spent millions of dollars to buy land, and it's not just on the Cape. They will be seen from Martha's Vineyard and from Nantucket, and then they say we don't want this, and then they say [another snigger] this will hurt seabirds. I don't buy that. I think it's more of a view thing myself."

Carr next asked Roosevelt for his thoughts.

"We have a gorgeous piece of land there that we got for a song twenty years ago," the investment banker replied. "When this goes up we'll see it. You'll see probably the red lights. You'll probably see the turbines, etc., but in my opinion, this is a very small price to pay to begin the important steps in getting this country energy independent and addressing the question of global warming, which I think is arguably the most important environmental issue we'll face."

Gordon couldn't believe his good luck. Where had this guy come from?

"How does your wife feel about this?" Carr asked.

"She *hates* my position," Roosevelt answered candidly. "She and I have bitter arguments about this." He added that his wife preferred that the area be a national park.

"This Cape, Martha's Vineyard, Nantucket—it's a glorious area and she thinks this is going to be a way to just utterly trash it. She hasn't looked at the whole picture. She's just not up to date," he continued. "Her reaction is the reaction a lot of people have."

Corrigan still hadn't shown up. Carr, saying he was speaking for Corrigan, accused Gordon of trying to make a lot of money by building the project.

"I hope," said Gordon, "this is a profitable project. I want to show that it's not only polluters who profit" from building energy projects.

Roosevelt interrupted Gordon.

"It would be *terrific* if he made money on it because that's going to mean that more money will go into wind farms, and we would then start to have alternative sources of energy. Right now in this country, 80 percent of our energy comes from burning fossil fuel, burning oil or coal. *Way too high!*"

Theodore Roosevelt IV speaks about the importance of wind energy during a press conference at the National Press Club in Washington, D.C.

Cape Wind staff, listening in their offices, were blown away. Not only did Roosevelt, with whom they had never spoken, support Cape Wind, but he had an impressive array of facts with which to back up his support.

Instead of talking absolutes, he talked about trade-offs. He was pragmatically insightful, refreshingly rational.

Roosevelt's presence on a radio talk show was something of a fluke. It wasn't the world the great-grandson of the Rough Rider president typically inhabited. But Carr's station was owned by Entercom, one of the nation's largest radio broadcasters, headed by David Fields. Fields, a Roosevelt friend, had asked the investment banker to do the show. After agreeing, Roosevelt had researched the Cape Wind project by calling a number of his most trusted environmental organizations.

He had never met Gordon, but he had long been a wind-energy supporter, a fact consistent with the rest of his life. TR IV chaired the Pew Center on Global Climate Change, opposed drilling in the Arctic National Wildlife Refuge, was vice-chair of the Wilderness Society, and belonged to Republicans for Environmental Protection.

He was also a trustee of the American Museum of Natural History, founded by his great-great-grandfather, whose son, the future president, had begun bringing animals—both living and dead—into the family mansion. Family lore had it, said TR IV, that his great-great-grandfather was "driven to establish the Museum because the maids doing the family's laundry were in revolt having to dodge a large snapping turtle tied to the sink in the laundry room and his sisters were outraged by the smells coming from his upstairs 'museum' filled with his early taxidermy."

Like his presidential ancestor, this twenty-first-century Ted Roosevelt often spoke his mind. He also distrusted the oil interests, also saw life as a series of trade-offs, and also supported social justice issues. As a managing director of Lehman Brothers, TR IV had helped several emerging-energy-technology companies with imaginative ideas find capital. Moreover, long before the September 11 disaster, in talks with titles like "What Does Your Group Think About Wind Power?" he had spoken strongly about the importance of integrating wind power into the nation's energy mix.

And as for politicians reluctant to lead their nation into energy reform, he charged them with "a moral cowardice, a political cowardice." Real leadership, he felt, led people to *engage* with the important issues of the day. Like his great-grandfather before him, he believed that democracy depended on the genuine participation of the people, and not on their manipulation.

Like his ancestor, he scorned the concept of privilege. Once in 1910 in London, the Old Lion was told a king had come to visit him. "Confound these kings!" shouted the ex-president. "Can't they just leave me alone!" Roosevelt IV tends to have similar responses. On hearing Bunny Mellon's "traitor to your class" remark, he first looked shocked and

Wind opposition spokesman Ernie Corrigan (left) discusses Cape Wind with Jim Gordon (right).

then profoundly pained. "That kind of thing," he said, "will eventually hurt them."

After Corrigan finally arrived, the radio show deteriorated into the usual charges of Gordon being a "greedy" developer, of the project harming birds and other wildlife, and of the project ending in raising electricity rates. But Roosevelt added a welcome voice of reason, based on facts and on history. When Corrigan raised the issue the Alliance had had the most success with—public land used for private profit—Roosevelt spoke about historical precedent.

"You raise in my opinion I think a very good point," he said. "We have a question of environmental justice here. The people in fact who are most opposed to this are people who have houses on Nantucket or houses on Cape Cod or houses on Martha's Vineyard. And then we have a very polluting power plant in Sandwich. And the people who are downwind to that are subject to getting mercury, sulfur oxides, nitric oxides. This is clean air!"

"I think I see a question of environmental justice here, which has been ignored by Ernie to date," TR IV continued. "And I also think I see a question regarding the use of public land. Historically, the United States has always encouraged industries. When we were building the railroads in the nineteenth century, to get them going the federal government helped them. In the case of the railroads, what did we do? We gave them alternate square miles along the railroad bed." (The gift to the railroads was one reason that President Theodore Roosevelt felt justified in holding the railroad men accountable when they price-fixed their transportation rates.)

A caller raised the question of the electricity rates. Would the project, the caller wanted to know, lower electricity rates, or would it raise them? Again Roosevelt stepped in. "You've asked what I think is a very good question," he said authoritatively. "I don't think anybody today in this room can answer that question. And it's a very good one, it's a legitimate question. But we can probably get a pretty good idea of what the future is likely to hold. Most people in this room and I think most people in this United States recognize that the cost of natural gas per thousand cubic feet or a million cubic feet is going to start going up."

"We've seen it go up dramatically over the past year. It's going to fall back a little bit, but the fact of the matter is that there's a shortage today of natural gas versus the demand. The price of oil is probably going to continue to go up in real and in absolute terms."

"Here is something that we can, once it gets built and once you've got all the cost loaded into it, will be able to produce electricity at a fixed cost which in all probability, I can't say for sure, in all probability will be lower than where natural-gas prices will be five or six years from now."

Raising the issue of natural-gas supplies was prescient. Only a few days later, New England underwent a dangerous cold snap that was to test the vigor and reliability of its energy-delivery systems to a degree never before experienced. For almost a week, temperatures dipped below zero at night, and rose only to the single digits during the day. Newspapers called it the worst cold snap in twenty years. New England's electric-grid managers at the system's central offices in Holyoke, Massachusetts, were on high alert, working overtime to ensure that the lights stayed on, despite the region's bitter temperatures.

The region's power grid is a spiderweb-like network of 350 power plants spread across six states and connected by 8,000 miles of transmission lines, carrying power throughout the region, and thousands upon thousands of miles of distribution lines, which distribute power to individual consumers. The system brings electricity to 14 million residents via 6.5 million electricity meters. This colossal technological marvel, which has evolved over the past 100-plus years, is today managed by a not-for-profit corporation called an Independent System Operator (ISO), based in Holyoke, Massachusetts.

ISO load managers are the unseen and unsung heroes who keep New England's lights on. Their boss is a soft-spoken southern gentleman, Stephen G. Whitley, ISO's chief operating officer, a retired colonel in the U.S. Army Reserves who usually wears his Screaming Eagle pin on his lapel, and a man with more than thirty years' experience in managing electrical grids throughout the nation. Whitley is a friendly man, highly respected in the industry. Even power-plant owners who aren't always happy with the market implications of some of his decisions find the white-haired professional fair and even-handed.

If the lights go out, it's his head that will roll.

Whitley, however, is the kind of man who can handle that pressure. When the senior executive arrived at work the morning of January 14, and evaluated the situation on the first day of the cold snap, he checked out the control room's big board, a forty-seven-foot by twelve-foot electrically powered diagram of the complete network, monitored day and night by a highly specialized team of system load managers and forecasters.

The board showed Whitley at a glance how many power plants were online, delivering power. Checking with his load managers, he looked at predictions of how much electricity would be consumed that day, then checked how many plants promised to be available if needed.

This is a promise that becomes particularly important in an emergency, but Whitley, while attentive, wasn't overly worried. Roughly 10,000 megawatts of gas-fired power plants reported that they could serve the system if called up, in addition to the coal-fired, oil-fired, and nuclear-powered generation already online. On winter-peak days, the grid typically uses somewhere between 23,000 and 25,000 megawatts. With 10,000 megawatts of gas-fired power ready for action, Whitley was looking at what appeared to be a safe cushion.

But by 8:30 that morning, the relatively sanguine picture began to change. A trickle of phone calls began coming in to the Holyoke headquarters, all with pretty much the same bad news. Plant operators who relied on natural gas as their fuel reported that although their plants were in working order, there was no gas available for them to buy. It had all been taken by the companies responsible for providing gas for home heating.

By afternoon, the trickle of "no gas" calls became a flood.

Now Whitley was deeply worried. In the wintertime, electricity use peaked during the late afternoon and early evening, when lights and televisions were turned on and when ovens began cooking the evening meal. During this all-time winter peak, when electricity was essential for the very survival of many New Englanders, roughly 7,200 megawatts of gas-fired generation was now unavailable.

Among the plants reporting a supply problem was the natural-gas-fired power plant at Dighton, built years earlier by Gordon and Energy Management, Inc., then sold to a California-based energy company called Calpine.

"We're going to be offline," plant operators told ISO. "We can't get the gas." When Dighton's control-room operators had been asked to begin generating electricity, the operators had called their purchasing and marketing departments, asking for the gas, only to be told by the marketing department that the plant was unable to fulfill its responsibility because they couldn't find enough natural gas to buy.

Dighton's engineer Matt Palmer, then volunteer executive director of the pro–Cape Wind nonprofit Clean Power Now, was one of the Dighton crew forced to sit out the crisis. It was a terrible experience for Palmer and his colleagues. Dighton's staff felt helpless.

"No gas to be had for love or money," thought Palmer, who quickly realized the gravity of the situation. For a plant engineer, it's a point of pride to be able to get his plant up and running, to respond in an emergency, and to contribute to the stability of the system as a whole. During grid emergencies, engineers at all 350 power plants have a common mission that develops a strong sense of teamwork and camaraderie. Yet there they all were, sitting uselessly, not because they hadn't done their jobs, but because the fuel simply was not available. Because the same supply of natural gas was also used for home heating, the pipeline system simply couldn't keep up with the demands of both heating and electrical generation.

Palmer and others felt they were inadvertently letting their teammates down. Sitting at his control-room computer, Palmer didn't immediately make a direct connection between Gordon's project, the lack of natural gas, and the cold-snap emergency, but he did understand that the proposed wind project wouldn't depend on a vulnerable energy-delivery train, one that could be easily disrupted by terrorists or could otherwise be interfered with. Relying on fuels that had to be transported long distances, sometimes from different continents, he thought, was absurd. New England had its own indigenous power source—strong and dependable winds—and here were the Dighton engineers, responsible for their own roles in keeping New England's lights on, and all they could do was sit on their hands.

While Palmer and his colleagues fretted, weather forecasters predicted a nighttime low of minus ten degrees.

In Holyoke, despite the cold, Steve Whitley began sweating bullets. There are graduated steps laid out for such crises. Whitley immediately started down that road, putting out a call for electricity conservation. He had nonessential power use curtailed at some business locations that had agreed earlier to participate in that program. He ordered staff to look for emergency power in places like New Brunswick, but that was no easy task, since neighboring regions were also coping with the deep cold.

As the situation worsened, he opted for more severe steps. To maintain reliability, problems that might occur in distinct parts of the grid would need to be contained and isolated, quarantined in a sense, before those problems spread to infect the whole grid, bringing down the system throughout New England. It had happened before elsewhere in the nation, and Whitley was determined it would not happen under such dangerous circumstances. The system was already on a knife edge. If any currently generating plants malfunctioned—a strong possibility, given the effect of severe cold on electrical equipment—there would be no margin available, no way to make up the shortfall.

Finally, Whitley decided to prepare to go to OP7—Action in an Emergency—the last step in avoiding a systemwide catastrophe. The New England power grid had never before needed to take this action. But a grid crash, which had never before originated in New England, was not the kind of innovation Whitley wished to become known for.

By the time he put out the order to prepare for an OP7, Whitley had spent a considerable amount of time on the telephone, engaging governors, getting out conservation appeals to the region's businesses and households. All of this helped considerably, but Whitley wasn't taking any chances. He didn't want the region for which he was responsible to experience the 2003 fiasco in which an overheated wire in Ohio took out much of the northeastern electrical grid—except for New England, which had avoided the disaster due to its modern automation and good grid management.

As part of his order, on January 15, 2004 Whitley had New England's substations staffed manually. He wanted to be ready if temporary power shutoffs of about fifteen minutes at a time in various sections of the grid were required. This strategy is what the press calls "rolling blackouts." Such an emergency had never occurred in the region.

No one at ISO was getting much sleep. In Vermont, some equipment did malfunction. In New York State and in Quebec, electrical grids were also having problems due to the extreme cold, making their ability to pitch in and help New England rather doubtful.

Finally, by Friday, the roiling chaos began to settle. In Boston and elsewhere, schools were closed. Some businesses either shut down or went to minimal staffing. This substantially decreased the demand for electricity. The crisis over supply of natural gas slowly eased. More gas-fired plants could go online.

Then, perhaps most important, winter temperatures began to moderate.

Emergency staff were allowed to stand down.

"In the end, the system all held together—but just barely," said New England's ISO president, Gordon van Welie.

Out in Nantucket Sound, the wind often blows best when the weather is at its most extreme. During the cold snap, Cape Wind's wind-speed monitor on its data tower showed that the wind was indeed blowing at speeds optimal for electrical generation.

The crisis caused some regional energy experts to take a second look at Jim Gordon's proposal, and at wind energy in general. When wind turbines produce electricity, fewer fossil-fuel generators are called upon by grid operators. In industry parlance, they are "backed down." Using new wind cacpacity, out in the Midwest, Xcel Energy, an eleven-state utility, had even successfully backed down old-fashioned coal-fired plants, a feat that many wind-technology opponents had claimed could never be done.

During the 2004 cold snap, "the Cape Wind project, if it had been fully constructed and online, would have made a significant contribution to the power supply and reliability of the regional grid," said a Department of Energy white paper. "Over the three days, the project would have delivered 25,596 megawatt-hours of power and would have averaged 396 megawatts per hour."

ISO executives did not make a specific statement regarding Cape Wind. Responsible for synchronizing the region's privately owned power plants but not directly responsible for siting or operating those individual plants, ISO takes no stands on individual projects.

But van Welie has repeatedly spoken to gatherings of the region's governors, to industry experts, and to electricity consumers about the importance of adding generation and of improving the mix of types of generation while planning for the future of the region's electrical power. "Given that New England's demand for electricity is growing at around 600 megawatts per year, and given that as a general matter it is very difficult to site non-gas-fired power plants in New England, it is likely that we will continue to add a significant amount of natural-gas-fired generation to the New England system, unless the region makes it easier to site nongas alternatives," warned van Welie. Decisions made today will impact New England's electrical grid ten years from now, he continued. "We are going to drive our costs through the roof, then wonder what happened."

ISO executives did, however, create an emergency task force after the fact to study how such a serious situation could be avoided in the future. The second time around, they worried, the outcome might not be as benign. The fuel-diversification task force focused on the possibility—even the likelihood—of the natural-gas shortage recurring. Without power-plant diversification, the task force made clear, the next time, the power could go out. An ISO statement said that New England had an "urgent need" for new generation and for a diversification of its energy mix, in order to keep a crisis like the cold-snap event from recurring.

On that task force was the DoE's Albert Benson, the behind-the-scenes expert who had been continuing to try to explain that building projects like Cape Wind could help keep people safe in emergencies and could also reduce energy costs throughout the region by offsetting the pressure on natural-gas supplies.

"The cold-snap emergency clearly demonstrated that wind can play a very important part in supporting the electrical grid during winter peaks," Benson said later. "Further reliance on natural gas to supply the region's electricity does not appear to be the least-risk course of action. Large additions of wind generation are needed to increase system reliability. Wherever it can be put in—Marblehead, off Cape Cod, both onshore and offshore, it's important. But the problem is that because of the lack of land, it's hard to find places where you can put in hundreds of megawatts. Offshore locations seem optimal."

Since 2001, Cape Wind opponents like the Cape Cod Chamber of Commerce's John O'Brien and the Alliance to Protect Nantucket Sound had been telling Cape Codders, Massachusetts residents, and New Englanders at large that the region would need no new power plants for many years. The facts were at odds with the Alliance's misinformation.

Nevertheless, ISO's frank and honest statements of the true situation offended Alliance funders. The Alliance's attorney dragged ISO executives into legal proceedings, ultimately claiming the grid managers were "lobbying" for Cape Wind.

Since the expenses for operating the grid are rolled into rate-payers' monthly bills—the average New England homeowner pays, very roughly, about sixty cents a month for the Independent System Operator—the cost of the Alliance-engendered legal proceedings against the ISO would, of course, also be passed on to New England electricity consumers.

But perhaps more important, the Alliance's tactics of intimidation would further discourage power-plant developers from trying to build the much-needed generation that kept New England's lights on and its people safe and warm in the wintertime. "A grid outage during a major cold snap will not only cause economic dislocations, but serious loss of life," warned Benson. "People could die because of this. It's not funny."

Watching the situation closely, many people began to believe that the funders of the Alliance to Protect Nantucket Sound—and the politicians following their lead—were playing with fire.

CHAPTER TWELVE

The Pleasure of the Governor

Democracy is only a dream; it should be put in the same category as Arcadia, Santa Claus and Heaven.

H. L. Mencken

Massachusetts governor Willard Mitt Romney dealt with the cold-snap emergency by handing out blankets to Boston's homeless. Some wondered if *that* was the governor's energy policy. If so, it seemed rather Dickensian.

Jokes made the rounds: What's Governor Romney's energy policy?—"It's cold here boys. Let's head for Palm Beach." Or, What's Governor Romney's energy policy?—"Just Say No!"

For climate- and air-quality activists, the early days of Romney's administration had been halcyon days, days of hope and of hard work looking for energy solutions that many believed might actually be implemented. During the campaign, Romney had promised to bring the state's dirtiest power plants, the "Filthy Five," to heel.

Advocates took heart, particularly when Romney hired Doug Foy, the aggressive and sharp-spoken head of the Conservation Law Foundation. Since 1975, Foy, a Princeton physics graduate, had pushed his organization into the fore of New England environmental activism. An impatient man famous for getting what he wanted—supposedly he had sued the state of Massachusetts more than anyone else in history—Foy thought Romney's offer was too good to pass up. Far from feeling coopted, Foy believed he would have an opportunity to make things happen. Environmental advocates lauded Foy's appointment, believing that Foy, coming from outside the political world, would look at an issue and make a call on its merits. When that happens, activists believed, the decision usually favors the environment.

At first Romney did seem to live up to his campaign promises. He spoke publicly about air pollution. He pressed polluting power-plant owners to install scrubbers. Once, when employees at a coal plant infamous for its severe pollution events loudly heckled Romney, the governor yelled back. "I will not protect jobs that kill people," the *Boston Globe* reported Romney shouting. "And that plant kills people."

Advocates' jaws dropped. Among those listening to this exchange was energy activist Frank Gorke, of the Massachusetts Public Interest Research Group. Standing to the side of the audience, Gorke was stunned. There was the Massachusetts governor, speaking from his portable podium, flanked by several well-known clean-energy advocates. Behind the governor, Gorke saw the smokestacks of the Salem Harbor Station power plant. It was a powerful scene.

And only twenty or thirty yards away, the mayor of Salem, a very angry man just then, was preparing his own press conference. The governor's staff had not told the mayor about Romney's appearance in Salem, an egregious political discourtesy that had greatly annoyed the mayor, who, in any case, opposed the governor's clean-air restrictions. The regulations, the mayor said, would cost his city jobs.

Standing at their podiums, each conducting press conferences, the two politicians looked like sideshow barkers, trying to lure customers away from the competition, into their own tents.

"Bizarre," thought Gorke, admiring the governor's courage. "It was a politically fraught situation, but he's standing by the facts. He's following through on his campaign promises."

When all was said and done, when the podiums were packed up and the television cameras gone, the governor got into his long black limo. He turned to Foy. "That was fun," Romney told Foy. "Let's do it again sometime."

Foy passed that remark around Boston. Advocates celebrated.

That kind of fearless, no-nonsense talk led people to believe they had elected a genuine leader, a strong-minded man who would not be buffeted by political winds. The role suited Romney perfectly. Throughout his business career he had been a troubleshooter and a change agent. He very much enjoyed his reputation as a hero and man who could get things done.

While shouting back at the hecklers, the governor had pointed toward the Salem power plant, a 750-megawatt dinosaur and the subject of an ongoing and bitter controversy. The plant had been burning coal since 1952—the year a London fog killed 12,000 people. The Londoners had died when coal particulates held at ground level by one of the city's famous fogs acidified the air, which then ate away Londoners' lungs. Over the past fifty years, the Salem plant had had lots of "fixes" but problems remained. In 2001, Republican governor Jane Swift had issued very strong clean-air regulations for Salem and several other plants. Nevertheless, the coal burner was likely to remain essential for years to come, since New Englanders' reluctance to build new generation while also insisting on low electric rates had made the cheap power indispensable.

Romney's finger-pointing may have been caused by a similar controversy in his favored state of Utah. In the late 1980s, a Brigham Young University scientist found that when a mill operated by the Geneva Steel Corporation temporarily shut down, two-thirds fewer children were hospitalized for breathing problems. Armed with this powerful data, outraged Utah residents fought hard to limit the mill's emissions. It was a very big fight. Everyone in Utah knew about it.

If advocates thrilled to Romney's behavior at Salem, they were happier still after Romney joined an April 2003 initiative begun by New York Republican governor George Pataki. He had invited eleven northeast states to join together to create a kind of mini Kyoto Treaty, the Regional Greenhouse Gas Initiative. "Now is the time to take action towards climate protection," Romney wrote Pataki. Meanwhile, Romney aides were beginning to develop a state plan, the Massachusetts Climate Protection Plan, which would eventually include seventy-seven measures intended to improve air quality, improve energy efficiency, and even lower the cost of energy. Activists crossed their fingers, allowing themselves to believe, maybe just a little bit, that the state was entering a New and Shining Age. "Maybe Massachusetts really will lead the way," mused Gorke.

But in 2004, things fell apart. The January cold snap made clear to New England's six governors that the region faced serious trouble. New England's electricity consumers had survived the cold snap by the skin of their teeth, ISO officials warned. What about next time? The region desperately needed new generation, including lots of renewable projects. Yet Romney took no action to encourage new power plants. Under his administration, a few solar panels had gone up. A few buildings were made energy efficient. Other symbolic actions occurred, but no real change took root and grew. Moreover, the governor was backing down from his earlier, bold, finger-pointing leadership. He gradually began to talk about easing emissions restrictions.

Even worse, when he unveiled his Massachusetts Climate Protection Plan, which acknowledged the climate-change crisis, the governor attached a signing statement denying his personal acceptance of the science. Foy later explained that the governor was "agnostic" on the climate-change debate. Romney's 2003 fans felt betrayed by the waffling. The *Boston Globe* headlined the story with "Romney Hedges on Global Warming." Clean-air advocates were saddened. "If he does the right thing, that's ok—for now," thought Gorke, who suspended judgment. Later, though, Romney pulled out of the Pataki-inspired regional climate-change agreement. "He flip-flopped," Gorke then thought. "He chose to be on the wrong side of history."

Simultaneously, Romney's efforts to stop Cape Wind escalated. His earlier somewhat low-key opposition burgeoned into a veritable crusade. Many people wondered at the governor's vehemence. Why didn't Romney get it? He was missing an important political opportunity. Leading the charge in support of Cape Wind would position him as a leader with both the right and the left. He could have shown himself both a realistic environmentalist and a promoter of visionary entrepreneurialism, appealing particularly to Republicans. Instead, Romney made himself into an example of John Bullard's theory of NIMTOF: Not In My Term Of Office.

By 2004, many voters realized that Romney was using the state's gubernatorial office as a stepping stone to his real goal, the American presidency. The more Romney focused on the Oval Office, the greater grew his Cape Wind opposition. Perhaps still thinking of Gordon's proposal as a little local power-plant battle, he failed to grasp the project's emerging status as a national symbol of energy activism. Americans wanted action. Instead, Romney continued to shoot himself in the foot by chanting Alliance mantras, by buying into the saccharine sentiment that Nantucket Sound was a "national treasure" somehow comparable to Yosemite or Yellowstone. Away from Cape Cod, those Alliance-originated sound bites rang hollow.

Fearful for their candidate's future, Romney's staff begged him to stop. If he didn't want to change his stand on Cape Wind, they said, just keep quiet on the subject. The inability to revisit a decision worried them. It could be seen as a fatal flaw in a man who wanted to lead the Free World.

But the future presidential candidate refused to listen to his advisors. The standoff had the distinct air of machismo. Romney increasingly flexed his political muscle. He felt he had to deliver. Why would his financial backers believe he could help them from the Oval Office if he couldn't help them as governor?

At fifty-five and still trim, Willard Mitt Romney had the kind of square-jawed good looks that guaranteed he would appeal to a certain type of woman. Indeed, everything about his six-foot-two-inch physique promised firmness, masculinity, and a Christopher Reeve–like heroism. "A real gift for the tube," commented one man who found the governor's appearance attractive. Some journalists swooned over this virile image. In the few in-depth pieces written during his early political career, there were bits about office secretaries turning their heads when this self-styled savior-of-corporations-in-trouble walked by.

On the other hand, Romney's political opponents had nicknamed the governor, with his *always* carefully coiffeured appearance, "Ken Doll." And it was true that the man's appearance was almost eerie, in that he never was seen with a hair out of place, making him the envy of all women who suffer bad-hair days.

If by 2004 his speeches had become a bit like his hair—he could certainly no longer be counted on to point his finger—the governor's massive intelligence was also obvious. The youngest of four children, Willard Mitt (the first name comes from J. Willard Marriott, founder of the hotel chain, fellow Mormon, and close friend of Mitt's parents) was a hard, hard worker who had proved himself as such since his earliest days of adulthood. The son of George W. Romney, who had saved the American Motors Corporation from collapse during the 1950s and then became a three-time Michigan governor, Willard Mitt led a childhood of economic privilege. He could have chosen to live the undisciplined life of a Ted Kennedy or a youthful George Bush.

He did not. Instead, he chose a life of rectitude that was almost nineteenth century in its correctness. Born in 1947, he married his high school sweetheart. He was valedictorian of his class at Brigham Young University. He graduated cum laude from Harvard Law School and in the top 5 percent from the Harvard Business School. No amount of money can buy an academic record like that. Not even genius can bring that kind of success. It takes work, study, and discipline.

Part of his discipline derived from his devout Mormonism. Also known as the Church of Jesus Christ of Latter Day Saints, the Mormon church is ruled by a small group of men who don't drink, don't smoke, and value financial success. Mormon men are confident men. They believe strongly in the rightness of their ideals, and, true to their custom of sending their children abroad to preach their gospel, they firmly believe their ideas are right for everyone. This Mormon self-assurance brings great wealth for many, and, for some, political power.

After a few semesters in college in California, Romney decided to follow the traditional path and go overseas to proselytize—although it must be said that he did not, like many Mormons, choose a developing country, but instead chose France where his father's friend Sargent

Shriver, then U.S. Ambassador to France and a Kennedy in-law, could look after him. Shriver's help turned out to be invaluable. Toward the end of Romney's tour of missionary duty, he was at the wheel with four others in a car when another car ran into his. One passenger in Romney's vehicle died. The accident was ruled the other driver's fault. Shriver helped the Romney family cope with the crisis.

After France, Romney finished his schooling, graduating in 1975 with his Harvard degrees, then took a job with the Boston Consulting Group. He jumped ship a few years later, when entrepreneur Bill Bain asked him to join Bain & Company, dubbed by *Globe* columnist Alex Beam "a cult-like consulting firm." Beginning in 1978, Romney became vice president of that secretive and wildly successful company.

In 1984, Bain again tapped Romney, this time to head Bain Capital Inc., a new private-equity venture capital and leveraged buyout firm. In that capacity, Romney proved himself a tough chief, and sometimes, a competent hatchet man. In the first year he raised $37 million from wealthy individuals. He also made hard-nosed decisions like laying off employees. Led by Romney, Bain Capital invested in a number of successful companies like Staples, the discount big-box office supply chain, and Domino's Pizza.

As head of Bain Capital, Mitt Romney became a very wealthy man. In the early 1990s, Bain & Company went on the skids because of fiscal overextension. Heading up the turnaround, Romney became a corporate cheerleader for the distressed firm, appearing frequently in the *Boston Globe* and the *Boston Herald*. As Bain & Company improved, Romney was increasingly thrust into the public arena.

Quite possibly, Willard Mitt Romney intended from childhood to run for the presidency. In a discussion with Sridharr Pappu for the *Atlantic Monthly*, Romney explained that his father had told him not to get involved in public life "until it is a question of service rather than employment." Romney, whose life pattern eerily parallels his father's, apparently agreed. The concept—fortune first, public office later—separated him from elected officials like Matt Patrick, whose lifelong vocation was public service, albeit substantially underpaid.

In 1994 Romney challenged Ted Kennedy. Apparently he had decided it was time for "service" in lieu of "employment." He had no experience as an elected official, although he had been president of the Boston Stake of the Church of Jesus Christ of Latter Day Saints since 1986, a position which gave him authority over eleven Mormon wards in the Greater Boston area. He announced his senatorial candidacy at Boston's Copley Plaza Hotel, where his famous father introduced him.

Much of his senatorial campaign financing came in four-figure checks from the state's wealthy residents. In a story headlined "Boston Businessman Top GOP Fundraiser," *Globe* reporter Scot Lehigh wrote that Romney had raised $256,000 in 1993 for his primary run, more than ten times the amount raised by his nearest Republican primary challenger. This was not surprising, given the high-finance crowd he ran with, but what was surprising was that most of his 2003 donations came from a "one-week flurry of fund-raising activity in late December," implying that his decision to run was last-minute.

The year of Romney's senatorial candidacy was also the year when the Richard J. Egan family came of age as political operatives. In a story headlined "Year of the Rich Guy: EMC chairman rose from a Dorchester paperboy to a major political contributor," *Boston Globe* writer Chris Reidy called Egan "Garfield on steroids," and enumerated the Egan clan and EMC Corporation's many political contributions. When all was said and done, more than twenty checks with the name "Egan" went into Romney's senatorial campaign coffers.

Romney lost to Kennedy but made a very strong showing, gaining influence and power in Massachusetts Republican political circles. Clearly, the 1994 election was not to be his last race. Also clear was that the tie between Romney and the Egan family would deepen as Romney gained political power. By decade's end, Romney had become Massachusetts's Republican Party leader. Dick Egan was Daddy Warbucks. Working as a team, they attended the same fund-raisings and the same political outings. They decided together who would and wouldn't run on the Republican ticket in Massachusetts. Together they made Paul Cellucci governor in 1998. The state Republican "Victory '98" election party was thrown by Egan and Romney, with bigwigs in attendance, like the powerful Andrew Card, future presidential chief of staff.

Romney took a detour back to Utah for several years to rescue the scandal-plagued 2002 Winter Olympics, after which the superman returned to Massachusetts well caped and covered in glory. So swathed, he announced his candidacy for governor for the 2002 election. Although he hadn't actually lived in the state for several years, he had maintained the necessary ties. The Egan clan was still on his party list.

Given this background, Romney's fervid opposition to Cape Wind is less surprising. After all, on the board of directors of the Alliance to Protect Nantucket Sound sat two of Richard J. Egan's sons, both of whom were Nantucket Sound waterfront residents.

In the political arena, pressure can take so many different forms. Sometimes the pressure is public, as with the anti–Cape Wind letters, speeches, and press releases from Romney, Kennedy, and others. At other times, the pressure is applied behind closed doors, making it all the more effective. So it came to pass that by 2004, in opposing Cape Wind, Romney had his fingers in several bureaucratic pies.

One closed-door tactic focused on an obscure state regulatory body called the Energy Facilities Siting Board. The name alone is enough to make a reader yawn. Few reporters attended siting board meetings; fewer still understood the board's importance.

This made it the perfect venue. The board was all that remained of Massachusetts's once-mighty regulatory hold over the state's utilities. State regulators once had immense input into the whole system of electrical generation. But once deregulation took effect, that power was greatly lessened. The state no longer determined electric-power supply rates. It could, however, in the case of Cape Wind, determine the placement of power *cables*.

Supposedly, the reason for this arcane curiosity was that, while power-plant construction had been left to the free market, the transmission and distribution system that brings electricity to consumers remained a monopoly. Therefore, went the theory, the wires must be regulated. Consequently, an independent power developer still had to jump through regulatory hoops, first proving a need for the cables (and thus, by implication, the project), then showing that the proposed cables provided both public benefit and minimal environmental harm.

Thus, in reality, the obscure board, rarely covered by journalists, could choose to wield quite a bit of power. Particularly important was the fact that a siting board permit superseded decisions from any other board or agency within the state, including, to the dismay of Cape Wind opponents, any town-level bodies. Because of its power to make legally binding decisions, the board was adjudicatory and *supposed* to be isolated from political influence. This goal was difficult to achieve, in that the governor appointed some of the board members. To protect them from the possibility of political influence, these appointees were supposed to receive three-year appointments.

To stop Cape Wind, project opponents needed to convince the Siting Board to deny Gordon permission to lay the project's two eighteen-mile power cables running from the project through state waters to their land connection. But this would be awkward, since the board had just permitted a thirty-three-mile cable bringing power through Nantucket Sound to ultrawealthy Nantucket Island, setting a precedent.

Fortunately, opponents didn't need to *stop* the project. They only needed to delay a decision until Gordon's patience wore thin, until he blinked and folded up, or until his investors backed out. Postponements kill electrical projects. Uncertainty terrifies financiers. The FUD factor—fear, uncertainty, and doubt—is very powerful. Investors who write checks during a power-plant project's development phase panic easily. The longer a project takes to permit, the more uneasy the money becomes.

When met with the kind of shock and awe manufactured by the Alliance and its politicians, moneymen wake up wondering what the future will bring. Too many mornings like that, and their enthusiasm is shot. They withdraw. Permitting the Nantucket Island power cable had required ten months. If opponents could stretch Cape Wind's permitting out for several years, the uncertainty would sow the seeds of discontent among Cape Wind's financiers.

But this strategy had a serious flaw: Gordon had no need to worry about sleepless investors because he *had* no investors. A few associates had chipped in several millions. The rest, edging toward $20 million, was Gordon's money entirely. And he was not a man who gave up easily.

And so began a series of eyebrow-raising delays that stretched first into months and then into years. Alliance lawyers peppered the board with a mass of requests. The Nantucket Island cable required two days of public hearings; Cape Wind's, twenty-one days. Siting board chairman Paul Afonso failed to schedule meetings for months at a time, apparently due to vacation schedules and other pressing matters. Typically, the board's professional staff—who have attended all the hearings involved in a case, listened to arguments from attorneys for both sides, and read the appropriate documents—issue a recommended decision. Then the politically appointed board votes on final approval only weeks later, usually affirming the judgment of its staff.

In the matter of Cape Wind, after the professional staff recommended approval, backed by a 190-page exhaustive discussion of evidence and arguments on both sides, the political board simply did not vote on the issue. The interim stretched to five months, then longer. Gordon's lawyer, David Rosenzweig, a siting board specialist, had never seen such delays. Gordon and others thought they saw the hand of Mitt Romney in these slowdowns. His interference was difficult to prove, but the governor certainly was active, despite the board's hands-off adjudicatory status.

In the summer of 2004, Romney focused on the rocks in Nantucket Sound. Specifically, the governor wondered which rocks dried out at low tide. "Drying rocks" poke above water at mean low tide. Incredibly, if such a rock lies within three nautical miles of a state, the rock legally belongs to that state. Setting the state's boundary, the rock extends the state's jurisdiction three miles farther out into the water from the rock itself.

Having found several previously unnoticed "drying rocks" in Nantucket Sound, Alliance financiers believed they'd achieved a coup. A newly drawn state boundary extended just a bit into the proposed Cape Wind site. *Et voila!* The state, they claimed, now had jurisdiction in the matter of siting Cape Wind. Gordon, they said, would have to begin the permitting process all over again. Romney was to raise this issue persistently over the coming year.

Gordon, ever the pragmatist, shrugged his shoulders. He would simply move the few affected turbines from the new state waters into federal waters, he told the press in early summer. Romney refused to drop the issue. Writing the siting board in late August, Romney said it "should . . . either delay any further review of the Cape Wind project until such time as the boundary issue can be resolved or should initiate a separate review of those portions of the project that would fall within state jurisdiction." The letter was unusual. Since the siting board was adjudicatory, most governors would not want to be seen as meddling.

Afonso finally scheduled a vote for November 30, 2004, twenty-six months after Gordon had applied for his permit. Gordon was eager to get the matter settled, but Alliance lawyers had another card to play.

The Army Corps' just-released 4,000-page draft Environmental Impact Statement *had* to be included in the record, the Alliance demanded. Moreover, they continued, everyone concerned would have to *read and digest* its information. More time was needed.

Deirdre Manning spoke up. Apparently, she'd had enough. The consumer protection commissioner for the state's Department of Telecommunications and Energy and, as such, a siting board member, Manning's whole career had been devoted to advocating for consumers. Her specialty was electricity. Her position was precarious.

Appointed by the previous governor, Manning's three-year term expired while the siting board was considering the Cape Wind matter. As Cape Wind dragged on, Manning waited for the governor's decision on reappointment. For nearly a year, she had been serving at the governor's pleasure. For some, this situation would make them eager to do the governor's bidding. But this Sword of Damocles did not stop the consumer affairs commissioner, whose salary was about $100,000 a year.

"The opposition to this project is the worst form of NIMBYism. The transmission lines should be approved today," she said. The press reported her remarks. Days later, Manning lost her job.

Romney replaced Manning with Judith Judson, a youngish woman with a Harvard MBA degree. She had worked briefly in the Romney administration and had then run for state representative against an incumbent Democrat. After losing that race, she needed a job. She had no work experience in the energy field.

The delays infuriated state legislators. Several committee chairmen fired off indignant letters. The siting board's stalling, said one letter, "raises serious questions about the willingness of the Board to carry out critical public policy mandates." Another letter said the board was "in clear violation of the well-defined timelines" laid out in the law. State representative Dan Bosley, a prime architect of the state's deregulation laws, complained that Romney was doing everything he could "to stall and delay."

Their words were ignored. With only Manning in opposition, the board had voted *not* to vote on November 30. Six months passed by. A vote was scheduled for May 2005.

As usual, Alliance lawyers began the meeting by calling for a postponement. The crucial issue this time: drying rocks. Pulling the nine-month-old Romney letter out of the stack, Alliance attorneys insisted that the rocks poking above the water at low tide made a completely new review absolutely essential. Back to the beginning for Cape Wind, they demanded.

Gordon's attorney countered that Cape Wind would move the turbines. As a condition of granting the permit, he said, the board could clarify that no turbines could be built in state waters, a solution that's typical in permitting matters.

Alliance attorneys insisted that Gordon's promise could not be taken for granted. More review will be needed, they said. Siting board chairman Afonso considered agreeing with the Alliance.

Gordon was furious. He saw Afonso and the Alliance lawyers pushing the matter well into late fall. He whispered—loudly—to his lawyers, "They know we're not putting turbines in state waters," in an uncharacteristic public display of frustration.

In a moment of high drama, particularly for the normally somnolent Energy Facilities Siting Board, Gordon grabbed some lined notebook paper. "To the Mass. Energy Facility [sic] Siting Board," he scrawled. "Cape Wind agrees that it will not build any wind turbines in Massachusetts state waters." Signing his name, he handed it to his attorney.

"Give it to him," Gordon growled, looking at Afonso. Rosenzweig chuckled, then planned his own little moment of high drama. He held Gordon's statement back for several minutes, letting Afonso continue to talk. Then, at the last minute, like Perry Mason, Rosenzweig dramatically handed Gordon's paper over to the siting board chairman. Afonso had no exit strategy. He could no longer delay voting.

He polled the other six board members. Four voted "yes." Only two voted "no": Romney appointee Judith Judson and Deborah Shufrin, representing Romney's secretary of economic development. Next, Afonso voted.

"Yes," he said.

Many in the room were surprised, but not Cape Wind staff. There was no legal basis for voting "no." Judson and Shufrin, however, looked thoroughly shocked, as though they'd been betrayed, hung out to swing in the wind, all alone.

From time to time during the Cape Wind War, several board members of the Massachusetts Technology Collaborative had tried to raise the issue of Gordon's project. The collaborative oversaw the state's huge Renewable Energy Trust, which at one time topped $150 million. The money, which came from the state's electricity consumers, was supposed to be used to develop renewable energy. Consequently, the board's interest in Cape Wind seemed well founded.

The collaborative had funded meetings on Cape Cod that were supposed to help residents sort out the complex project's pros and cons and cut through the fog of confusion and misinformation confronting the public, but since the *Cape Cod Times* rarely covered those meetings, the public knew little about them. The fund had also provided $300,000 for a survey by the Massachusetts Audubon Society of birds in Nantucket Sound. Otherwise, the collaborative was strangely silent.

After Manning lost her job, collaborative members again tried to address the matter of Cape Wind. Romney's collaborative representative, Carlo DeSantis, immediately moved to curb the uprising. Board members, some of whom were waiting for their reappointments, heard DeSantis warn: You saw what happened to Deirdre Manning, didn't you?

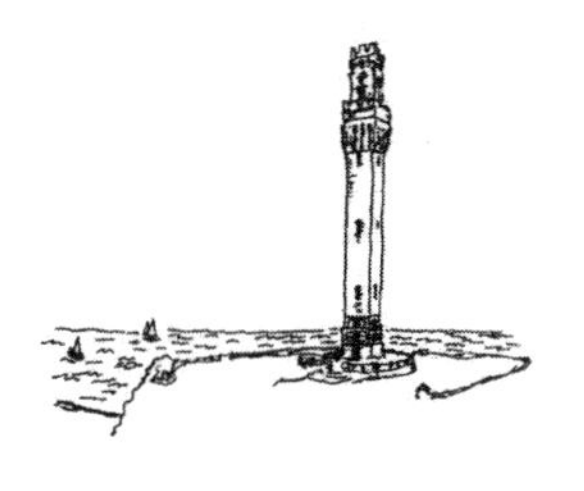

CHAPTER THIRTEEN

More "Public" Meetings

Those who expect to reap the blessings of freedom must, like men, undergo the fatigue of supporting it.

Thomas Paine

It was show time for Mitt Romney. Ten minutes late, the governor appeared at the second of the U.S. Army Corps' draft EIS public hearings. It was only one day after the Honorable Congressman William Delahunt had pounded his fist on the podium on Martha's Vineyard. But if Corps chief spokesman Larry Rosenberg thought the Alliance had run out of steam after ambushing the island's public hearing, he was sorely mistaken. All the usual Alliance suspects showed up for the second night running, armed with Christmas cookies and prepared to fight for control of whatever television cameras might arrive—which, thanks to the governor's presence, turned out to be quite a few.

Rosenberg had had only a day-ahead warning that the second night might be even more difficult than the Martha's Vineyard hearing. Just before the island hearing began, even as he was trying to fend off Alliance attempts to control the microphone, Rosenberg answered a reporter's phone call. The reporter told him the governor planned to speak at the West Yarmouth hearing the next day.

Rosenberg was floored. Setting up security for a governor was a major responsibility, and now he had only a few hours to get the job done. The state police detail had to be increased. An officer needed to be assigned to accompany Romney. The town police chief would need to be told. Extra town officers would need to be detailed. Which doors the governor would use to enter the school would need to be determined, then guarded. Emergency exits for the governor would need to be chosen. Then there were all the other little details. Which bathroom would the governor use? Was it clean? Did it have toilet paper?

Two of Romney's advance men appeared forty-five minutes before the governor. Rosenberg explained the protocol. As the state's top official, Romney would be allowed to stand at the stage podium to open the meeting. But, warned Rosenberg, the governor would need to be on time. Hundreds of people, possibly as many as a thousand, had come from all over New England, some just to listen, some to comment, some to report. This meant the hearing could continue into the postmidnight hours. The public had come to have their say, and it wasn't fair for the governor to make them wait.

The governor's behavior was unusual. Romney had already sent many letters to the Corps, including one claiming that the Corps couldn't move forward with public hearings because of the drying rocks. He had also sent a full statement commenting on the draft EIS, the subject of the night's hearing. In Rosenberg's twenty-plus years of managing public meetings for the military, a governor had never given him such late notice. In his fourteen years as New England communications chief for the Corps, a governor had never come in person to a public hearing. Typically, they send representatives.

The arrival of the governor was indeed exceptional, but Rosenberg had another surprise coming. Delahunt aide Mark Forest searched him out. "Attorney General Tom Reilly will be speaking for Congressman Delahunt tonight," Forest informed Rosenberg.

"No he won't," Rosenberg countered.

"Yes he will," Forest insisted.

Reilly did not have an official connection with Delahunt, Rosenberg said. Therefore, he could not represent him. Essentially Reilly, as attorney general, had no dog in this fight. He could sign up and speak from the floor microphone for the two minutes allotted to any other citizen. There was no protocol granting an attorney general special privileges.

Forest, seriously miffed, walked away in a huff.

Seeing Romney come to the podium, some in the audience groaned. Last night's speech by Delahunt had taken up so much time. Now the governor was doing the same thing.

Ernie Corrigan, however, was beaming. This was the show of shows he had long dreamed of creating. Corrigan had orchestrated many different ways of making the Alliance's presence known at the Yarmouth school, just as he had the prior evening. This time he had a large tent outlined with little Christmas lights placed outside the school, but directly in front of the public entrance. Under the tent, Alliance employees were handing out multicolored leaflets. Other Alliance staff, walking up and down the line of those waiting to sign in, handed out free coffee and cookies.

Looking over at the assembly of satellite trucks from the major New England television news outlets, Corrigan smiled. Everything was under control. The show *would* go on. He patrolled the halls, looking for opportunities for television face time.

Up at the podium, the governor began speaking. "This is citizen participation and I'm one of them," the governor said. He began with the usual, calling Nantucket Sound "a national treasure." Some in the audience yawned. The governor said he often enjoyed visiting the Sound himself and added that millions of people do so every year. Possibly the governor was unaware that much of the Nantucket Sound shoreline, while open to him as a guest of the property owner, was closed to the general public.

He continued by saying that he supported renewable energy. This made Cape Wind's communications officer, Mark Rodgers, chuckle. Rodgers had already won a bet with himself that whenever a speaker began with "I support renewable energy," the statement would slam Cape Wind. This was one sound bite that project opponents, for some reason, felt was very effective.

"I've seen wind farms," Romney told his audience. "They are not pretty."

"There are several areas in the Berkshire region where wind farms have recently been approved," the governor said helpfully. "There are a number of different areas along the coast of Massachusetts which may be appropriate as well."

A few listeners were genuinely shocked. Others looked quizzical. Was he saying that something ugly would be fine in the Berkshires, but not where wealthy people could see them? His remarks later earned him the headline "Cape Wind: Too Ugly for the Rich?" atop a column by *Boston Globe* columnist Joan Vennochi.

Romney listed all his favorite arguments, even the drying rocks. He closed by invoking the Deity. "This is not a decision about money," he campaigned. "It's not even a decision about power. It is a decision about our environment, about the legacy we leave our children. It is a heritage given to us by God. We may not, we cannot, trash this extraordinary resource . . . ."

If the governor was decidedly less flamboyant than Delahunt had been, his remarks nevertheless echoed those of the congressman and the Alliance. When the governor left the stage, public participation began. Rosenberg explained that since more than 400 people had signed up to speak, a traffic light would help speakers stay within the two-minute time limit. Microphones, he said, would be at the front of both auditorium aisles, and so on.

He was calling the attorney general as the first speaker when he felt a hand on his shoulder, pushing him away from the podium. The physical contact took him by surprise. Looking down at his notes and out at the audience, he hadn't seen or expected anyone to come up onto the stage.

He looked up and saw the attorney general. Rosenberg could do nothing but step back and let Reilly take over. In all his experience running meetings, no one had ever shoved him away from the microphone before.

When Clean Power Now chief Matt Palmer saw Reilly literally leap up onto the stage, his eyes widened. "Audacious," Palmer thought.

Palmer had been watching and learning over the course of the two evenings. This is not about public participation at all, Palmer thought to himself. This is about the guy with the most-est, who gets there the first-est, wins. Palmer knew his side had been blindsided at Martha's Vineyard. This time, he figured, Clean Power Now had at least fought the Alliance to a draw. Things would be different the next time, Palmer decided, right there and then.

As Reilly spoke, boos and catcalls came from the audience. "Get off the stage!" bellowed someone when the traffic light turned yellow. "Time's up!" cried someone when the traffic light turned red.

The public had been surprised by the governor's appearance, but generally courteous. But Reilly's assumption of the stage and podium seemed a bit much. It was widely understood that the attorney general intended to run for governor in the next election. Apparently Reilly wanted equal face time with Romney, who could very well be his future political opponent.

For the public, though, this was the straw that broke the camel's back. Seeing the size of the crowd at the meeting, people had settled in for a long evening, but they saw no reason to make it even longer just because the politicians were vying for votes.

Reilly's points basically paralleled those of the governor. "This is no wind farm," Reilly said. "It's a power plant."

The catcalls kept coming.

Gordon sat in the middle of the crowded auditorium, alone as usual, taking it all in. He had heard a rumor the previous day that Romney might come. When the governor did show up, Gordon just listened. By now, he was no political neophyte. He knew exactly whom this show was meant to impress—and, in fact, he *was* impressed. He understood that the governor's appearance made clear, as had nothing before, that Alliance financiers had political power to back up their money. The governor's appearance also made clear to Gordon that Alliance financiers meant business. They were willing to call in their chits with the governor of Massachusetts, a man who could possibly be the next American president, in order to get Gordon to go away.

As Gordon sat there listening first to Romney, then to Reilly, he was both disgusted and fearful for the future of his country. In none of his projects had he ever seen anything so blatant. This flexing of political muscle, meant to frighten him personally, did not bode well for the future of the state, of New England, or of the nation. "What's it going to take?" he wondered. "Romney is pandering to the Alliance, Reilly, to the Kennedys. They're just ignoring our energy-supply situation, our environmental problems, electricity costs . . . . The more he watched the politicians, the angrier he got. The politicians are showing they're good soldiers, he thought, but they're not doing their duty by the people in the commonwealth."

The more he thought about it, the steelier he became. Psychologists once discovered that repeatedly giving a caged dog an electric shock eventually caused the dog to lie down and give in to the pain, accepting its fate. Most dogs reacted that way—but not all. About 15 percent of dogs *always* throw themselves against the sides of the cage, *always* try to escape, *never* give in. "That's me," Gordon once said. "I'm still throwing myself against the sides of that cage."

As Reilly continued, Romney aides tapped the shoulders of television journalists and cameramen standing in the back of the hall. "The governor's going to have a press availability out in the lobby," the men said enticingly. "Come on out. You can ask the governor questions." Faced with a choice between filming the attorney general on the stage or having an opportunity to question the governor directly, most crews chose the "availability."

Perfectly groomed, cool as a cucumber, the governor motioned the TV crews to surround him. He said a few words, then opened the gathering to questions. First out of the gate was Kria Sakakeeny, a young reporter just out of journalism school. Unused to questioning men in power, she nevertheless stepped forward firmly, put her microphone up to the governor, and asked him if he weren't embarrassed to be taking up the public's opportunity for speaking out.

For an instant, the governor looked shocked. Then he composed himself to look hurt and saddened. "Well," he mumbled. "I'm also one of the people." He continued: "I know Senator Kennedy will want to speak . . . ." (Kennedy never spoke at these or any other Cape Wind public hearings.)

Fielding a softball from another reporter who was accommodatingly indirect, the governor was off and running about national treasures and the like. His aides smiled.

The politicians' presence at the hearing took up roughly an hour of time. Many of those signed up to speak gave up and went home. By the time the meeting ended, around midnight, the number of speakers who favored the wind farm roughly equaled the number opposed. But once again, few speakers addressed what was supposed to be the question at hand: the draft Environmental Impact Statement.

By the final public hearing a few days later, project supporters had decided to take matters in hand. Matt Palmer had had a few phone conversations of his own. Throughout the whole controversy, Palmer had managed to retain his faith in democracy. Clean Power Now had virtually no money—but it did have friends.

Cape Wind's Rachel Pachter stood shooting the breeze with Ernie Corrigan in the MIT building in Cambridge, Massachusetts. It was the last of the four public hearings. The long line of people waiting to sign up to speak stretched down the darkened corridor and around the corner.

Pachter heard chanting. "Save Our Sound! Save Our Sound!" The words came from around the corner. Pachter blanched and looked accusingly at Corrigan. "Don't look at me," Corrigan said. "I had nothing to do with this, but I wish I *had* thought of it."

Pachter had heard something fun might precede that evening's meeting, but she wasn't sure what it might be. Was this what Palmer was planning? From around the corner came the next line. "Save Our Sound! Save Our Sound! Especially the View from My Compound!" It was Corrigan's turn to blanch.

Five people dressed in old-fashioned yachting costumes, complete with the kind of little caps that Wianno Club members used to wear in the 1920s, rounded the corner carrying protest signs. The group began a new chant. "Cape Wind makes our Blue Blood Boil! Let's get our power from Middle East Oil!" shouted the yachtsmen.

One protestor carried a sign that said, Global Warming: A Longer Yachting Season. People standing in line looked confused. Another chant began: "Fighting windmills can't be hard! Just keep them out of my backyard!" As the group of five stood to the side of the hallway, newsprint and television journalists ambled over. No reporter can resist that kind of fun.

"OK, give me your name," growled a serious-sounding male reporter. "Spell your first and last names." Slowly the man in the white yachting cap removed a meerschaum pipe from his lips.

"It is Preston Cabot Peabody the Third," he said slowly, clearly and with very great dignity. He had binoculars around his neck and sported a white captain's shirt with black shoulder boards. "We're a new group," he said, in response to another question.

"We've formed as a grassroots organization . . . Hah! Grassroots! . . . " he chuckled, looking at another sign-carrier, ". . . to oppose these windmills. We feel that the windmills should be placed somewhere else. . . . *If* they're going to be there at all."

"We don't oppose these windmills," he continued, "but they should be in some place that is used to having power plants sited there, like Fall River . . . Roxbury . . . . We're also for flattening those mountaintops in Appalachia. I make millions off of coal mines . . . . I deserve to have my views. I've spent good money for my views. My captain's instructed me: It's very difficult to sail our yachts between those windmills . . . that's why we want them somewhere else."

"Yes! Yes!" a fellow demonstrator piped up, clapping loudly.

And so encouraged, the first speaker continued. "I think the answer to this is the Arctic National Wildlife Refuge. I think we should be focusing our attention on drilling in the Arctic."

At this point, a real SOS sign appeared behind Preston Peabody, a fact that Peabody would not dignify by noticing. It wasn't clear that everyone standing in line understood that this was a spoof. There are people in Osterville and Hyannisport who *do* still wear those caps and shirts.

"We want our views—and we want them NOW!" interrupted one of the group. Another voice among the five piped up. "We're not against energy. We believe that energy has to come from—somewhere else!"

"Appalachia," he continued. "What about the miners? What are they going to do?" He wore a porkpie hat, a shirt and tie, and a coat with a huge silk handkerchief stuffed into the breast pocket. "That seems a reasonable resource," he said reasonably. "Coal, nuclear power, Middle East oil. Those seem perfectly acceptable alternatives. Our point here—the reason we're here—is anywhere! Anywhere—but in our backyards."

Then, after a thoughtful second or two, he added another insight. "We just feel that it's . . . bad taste, among other things."

"Who's this guy?" the reporter asked Preston Peabody.

"This is my cousin Thornton," Peabody answered.

"My name," said the second man slowly, "is Thornton Prescott Beechcraft the Fourth . . . . You may have heard of the Beechcraft family. We're in the boating business. Pres and I are cousins. When we were out on the boat this summer, we said, you know, we've been spending

Yachtsmen Against Wind Now, a theater group that parodied Cape Wind's elitist opposition, make their voices heard at the fourth and final U.S. Army Corps public hearing to discuss Cape Wind's draft Environmental Impact Statement in December 2004.

millions of dollars trying to derail this windmill thing. We need to come out here and show our faces. And show our support for the people out here trying to stop this windmill thing."

He continued: "We are prepared to spend millions of dollars to drag out this regulatory process until this whole project can be killed. I mean, this Gordon chap! He should put his windmills somewhere else!"

To which the first man added: "Not in Nantucket Sound!"

"Maybe the Berkshires. The Berkshires is a good place!" said Beechcraft helpfully.

"They're not pretty," said Peabody agreeably. "But they're prettier in the Berkshires."

At one point during the public hearing, Ernie Corrigan rose to represent the Alliance point of view. "Over the past two years, the Alliance to Protect Nantucket Sound has conducted an education and grassroots campaign to inform the public about the real impacts of this project," the public-relations man said. "During the course of that campaign, we have gathered signatures from concerned citizens in the form of a petition . . . ." Corrigan said the petition contained names of people who were "opposed to the project." The signatures, he said, represented "the voices of the governed who are looking to you to protect their heartfelt opposition."

In the audience, the eyes of several members of Clean Power Now narrowed. The petition, again. This was getting to be like a vaudeville act. For months, this document had been a matter of quite a bit of ridicule. When it had first been aired several months earlier, some of the names had been, well, unusual. "Selfish Dingelberry," and "Craven Morehead," for example.

Clean Power Now's Jim Liedell had found something else unusual. At least one page of names appeared to be all in the same handwriting.

More interesting, Liedell thought, was that he knew several of the people on that page—and they did not oppose the wind farm.

A reporter called several of those people. One of them, Ann Rhuen, was surprised. "I don't have a position on the wind farm," she said, "because I haven't studied it enough." Another, Bob Churchill, said: "I'm not opposed to that project, and I never signed anything. My name should not be on that document. How it got there, I do not know." A third, Linda Campbell, said she opposed the project but did not sign the petition. Indeed, a dead giveaway to the truth, she said, was that on the petition her name was misspelled as "Cambell."

Nevertheless, entered by Corrigan representing "grassroots" opposition, these names are now part of the public record.

For Gordon, the hearing at MIT was a triumph, both professionally and personally. Many who took the time to come and to speak in favor of the project were people he'd never met. Unlike on Cape Cod, speaker after speaker said the innovative project should be given a chance.

Once we get away from the Alliance misinformation campaign and the carpet-bombing from the *Cape Cod Times,* Gordon thought with satisfaction, once we are away from that surreal atmosphere of Cape Cod, there are lots of people who support this project. All I have to do is find a way to get out of that spider web, he thought, and I can make this happen.

The enthusiasm of the meetings' speakers reinvigorated Gordon and the rest of the Cape Wind staff: There was something important they were trying to accomplish and they were no longer in this all alone. More and more people were standing up and speaking out.

Personally, the hearing was also special. Gordon usually went to great lengths to separate his family life from his professional life. He did not want his wife or son to sit through these meetings and be exposed to the character assassination that was so commonplace. Often, what was expressed was personal hatred, not just anger over the project, but very targeted remarks intended to paint him as a villain. For his own part, he didn't really mind. It's just business, he told himself repeatedly. But he didn't want his family to have to experience that kind of angst.

His wife, Meg, and his sixteen-year-old son, Zachary, however, had insisted on coming to MIT. Meg spent most of the time listening, holding her husband's hand. Zachary listened carefully. At one point, Gordon and Zachary were washing their hands in the men's room. "Dad," the teenager said to his father, "I just want you to know something. I'm really proud of you."

The crescendo of the Corps' public hearings was followed by a sharp and audible decrescendo. Perhaps the public was tired of theater, perhaps the winter was too cold for activity. For whatever reason, even the Alliance to Protect Nantucket Sound seemed to take a several-month siesta.

They appeared again, in grand fashion, in March of 2005. This time, once again, Mitt Romney led the way. The scene was Craigville Beach, "fourth best beach in the world" and, in the off season, a great place for photo ops.

Flocks of SUVs loitered like seagulls in the beach parking lot. Their drivers stamped restlessly in the chilly air. Boating weather was still several months away, and no one but Kennedys would be swimming in Nantucket Sound anytime soon.

The men were waiting for His Excellency the Governor to show up for his media opportunity. The air was fraught with tension, despite the "pristine" ocean setting. Many of those waiting had clearly decided that this wind-farm thing had gone far enough. It was time to put paid to the upstart Jim Gordon, who thought he could build a wind farm in their yachting territory. What's the point of donating to the governor if you can't get what you want?

Even their cars were angry. Stop the Steel Forest! demanded a Chevy Blazer (which gets about fifteen miles a gallon in stop-and-go traffic). Save Our Sound! ordered a Ford Explorer (another fifteen-miles-per-gallon vehicle). A Volvo, a 740 GL, called by at least one owner "a princely car fit for stately drivers," and a BMW 535i (listed as a "gas guzzler" on the Environmental Protection Agency's fuel-economy Web site) wore more anti–wind farm propaganda.

At long last, the governor appeared in a very long black car, flanked by a bevy of tall men in very long black coats. It seemed as though the pastels crowd, normally adherents to the green-slacks-with-little-blue-fish craze, had dressed for war. But to be honest, at least one foe did have on a lovely blue button-down shirt, bright yellow tie, and more-than-dapper golfing cap.

If the foes were dressed for war, Clean Power Now came dressed for fun. Tom Wineman, a CPN director who drove a ten-year-old Volkswagen (a fifty-miles-per-gallon car powered by biodiesel, which Wineman and buddies made by refining cooking oil from fish-fry restaurants), wore a six-foot-high wind turbine strapped to his back. He held his two hands high above his head. In them was a handwritten sign scrawled with felt-tipped pens. The sign had one word: YES. Ever enthusiastic, Wineman smiled widely. Other Clean Power Now members chanted and sang "The answer is blowing in the wind."

Ernie Corrigan, walking around the edges of the joyful mayhem, did not look a bit happy. The unflappable governor behaved as though the hecklers weren't there.

Topped off by super-well-combed hair with just a hint of respectable gray, the governor stood at his special governor's podium placed on the parking lot just at the edge of Nantucket Sound. To the west of the governor was the house of corporate mogul and famed wind-farm foe Doug Yearley. To the east was Squaw Island, Hyannisport, and the

Massachusetts's Republican governor Willard Mitt Romney takes to the seashore to expound upon his unwavering opposition to Cape Wind. Standing on Craigville Beach, Romney was located between the estates of the Kennedy family in Hyannisport and the home of Douglas Yearley, longtime head of the infamous Alliance to Protect Nantucket Sound.

Massachusetts governor Willard Mitt Romney explains his opposition to Cape Wind at Craigville Beach, while wind supporter Tom Wineman silently proclaims his support.

Kennedy Compound. Just over the governor's left shoulder was a sign on a fence that said: Private Beach. Behind the governor, across Nantucket Sound, invisible but always present, was Nantucket Island.

Out to sea, unnoticed by the darkly dressed men, a loon watched. The setting for the governor's declaration of intention to "protect" the state's oceans was more appropriate than the governor might have realized. Because of mercury emitted by fossil-fuel power plants, the watching loon, with its soothing cry that had come to symbolize natural New England, may see only half her progeny survive to adulthood.

The good governor commenced his hymn. "I'm all for renewable energy, but I don't think the right place to start is in a national treasure!" The loon seemed to groan.

"Nonsense!" cried a Clean Power Now member.

"I want my kids to breathe!" shouted another.

The governor's face time seemed to be deteriorating into an actual discussion, a free-for-all over the pros and cons of Jim Gordon's wind farm. This was not on the governor's agenda. His days of engaging with demonstrators were long gone.

The Alliance's executive director, Susan Nickerson, who had been standing winsomely at the governor's side, moved to the microphone. Perhaps not noticing the Private Beach sign behind her, she said: "This beach is publicly owned, accessible to all. It belongs to the public, the residents, and the taxpayers."

She was right . . . sort of. The very short stretch of beach that the governor had chosen was "public," in that the summer beachgoing public, other than town residents, had to pay twelve dollars per car on a weekend to enjoy it, or forty dollars a week. But other than the short stretch specifically designated for the public, most of this "fourth best beach in the world" was totally off-limits. Signs saying Private Property No Trespassing dot the landscape. Unlike other states, Massachusetts allows property owners to own the shoreline—right down to *low* tide.

This means that wealthy people who have always sought but never found that Golden Grail, a Beach of One's Own, can get just that in the land of the Puritan Fathers. It also means that the public is allowed to sit down and enjoy the sea only at places specifically designated for them. The state's wealthy take their beach ownership so seriously that a Nantucket Sound security guard once arrested the president of the Cape Cod Five Cents Savings Bank. Elliott Carr was strolling with his wife when three cars with flashing lights arrived. The elderly couple was arrested. "Ironically," wrote Carr with indignation, while the private-beach ordinances were enacted by the Puritans to enhance business, "they are now being used by private owners to prevent such activities. . . . Our shoreline is not user-friendly."

Elsewhere on the Sound, there are plenty of great beaches—if you have $130,000 to pay for a key to get through the beach gate. This Massachusetts law of exclusivity is one reason that so many wealthy people of a certain type have migrated to these beaches to build their huge mansions. The law is a major selling point. These property owners really want their views unobstructed—no wind turbines, and no people, either.

After all the speechifying had been seen to, the governor went down to the water's edge to see about a boat. He turned from the podium and set off at a run, trying to get there before the demonstrators could follow him.

The Alliance's fund-raiser, Pam Danforth (who earned $60,000 in regular compensation and $80,000 in deferred compensation for the 2003 tax year), staged a great photo op by rowing a little dinghy onto the beach. Governor Willard Mitt Romney bent over as though he were helping Danforth, and the photo ended up on the front page of the *Cape Cod Times,* with the cutline "Wind Farm opponent Pam Danforth of Centerville got a gubernatorial greeting."

Other photos were taken of the winsome Susan Nickerson standing with the governor. These photos were placed prominently in fund-raising literature, providing the clear but unstated message: "If you want access to Massachusetts's governor, contribute to the Alliance to Protect Nantucket Sound."

When the show was over, Romney's aides again tapped the shoulders of the gaggle of reporters following the governor, who had most graciously offered to provide a "press availability." This was supposed to be an opportunity for the press to ask questions, but the first query actually came from the governor himself. It was directed at the TV crews. "Am I standing all right for you guys?" he asked, offering up his finely chiseled square jaw. "Or is there too much glare?"

CHAPTER FOURTEEN

Capitol Enemies, Capitol Friends

Sentence first, verdict afterwards!

The Red Queen, *Alice's Adventures in Wonderland*

As the senator from Virginia entered the hearing room walking among his entourage, Dennis Duffy found himself fascinated. His eyes fell upon John Warner's handkerchief. That touch of southern elegance—a doubled-breasted dark suit set off by a small square of light-blue silk that matched his tie, complemented in turn by that famous leonine head of white hair—captivated the Cape Wind attorney. Breast-pocket flourishes like that were none too common in hard-boiled, Puritan-endowed New England. Duffy couldn't take his eyes off the senator's dapper livery. "Oh, man, this is going to be good," Duffy thought, chuckling to himself.

Duffy had come to Washington in late May 2005 to testify before the Senate Environment and Public Works Committee, which was investigating the difficulties of siting power plants. The hearing witness preceding Duffy had emphasized the need for organizing and streamlining the nation's siting process, which, to the witness, seemed to have evolved into a dangerously draconian, endless labyrinth of rules and regulations. Warner had missed that testimony. Perhaps he felt it was not germane.

But he did walk in just in time to put Cape Wind on the hot seat. Duffy was only just settling into his chair when he heard a soft singsong.

"If I could just have a few minutes with him . . . ." Warner requested gently of the hearing chairman. In spite of himself, Duffy smiled. He could only imagine what those "few minutes" would be like. He knew Warner was no friend of Cape Wind. Nevertheless, the Cape Wind attorney sat spellbound.

The U.S. Senate is not like any other place on earth. It was designed by the Founding Fathers to be an updated version of Britain's House of Lords, a body that had evolved over half a millennium to act as a counterweight to the sometimes-notoriously absolute authority of the British Crown. John Adams conceived of the American version, intending it to counter the power of the American presidency but also to curtail democracy's perceived tendency to mob rule, which Adams and some others believed would result if the House of Representatives, with its perceived close connection to the masses, were given full rein.

Indeed, the U.S. Senate was originally endowed with such semiroyal status that, for the first five years of its existence, it was a semisecret body. Casual observers were not allowed to attend sessions. No public notes were kept. And until 1913, when the Seventeenth Amendment was ratified, American voters did not have the right to elect their own senators. That duty was performed by the various state legislatures, who were thought, incredibly, to be more responsible than the average voter.

Things have changed since those early days, but vestigial traces of its original aristocratic structure remain apparent in the Senate even today, as is often observed by the more cynical among us. Senator John Warner, in office for more than thirty years and by many accounts an extremely honorable man, is one of those who continue to bear the burden of that certain *je ne sais quoi*, that vague sense of royalty coupled with an ancient dedication to noblesse oblige.

"I thank you for your courtesy," Warner lilted, after being introduced by the committee's then-chairman, James M. Inhofe, Republican from Oklahoma.

Duffy found himself really liking the guy. Up in Boston, if people were out to get you, they weren't so polite about it. Asked to testify, Duffy began with characteristic New England directness to make his opening statement. No sooner had he begun, though, than Warner interrupted the Cape Wind attorney's well-rehearsed presentation.

"Will you bring that mike a little closer and speak directly into it," the Virginia gentleman softly instructed. "I'm quite anxious to hear you. It's a directional thing," he said, moving on to explain to Duffy in most helpful and quite considerable detail how one uses a microphone. "He's trying to disconcert me," Duffy thought.

In recent months, Cape Wind had begun to develop a fairly high profile, and Duffy began by explaining why the project was so often in the national news. He briefly described a few of the project's technical details and a bit about Energy Management Inc.'s thirty-year history. He began explaining what the project could achieve in terms of renewable-energy development.

"The European projects have been operating for more than a decade now," Duffy explained. Warner interrupted again. "What . . . what has been operating . . . ?" Warner asked. Duffy repeated himself. "Riiiight. . . . Yes . . . ," Warner interrupted, then drifted off.

By now, Duffy, an experienced public speaker, was off his rhythm. He had only a few moments to speak, and he wanted to keep to his time limit. He returned to reading his statement, which, this time, Warner allowed him to finish.

The courtly senator began the questioning. "I won't be too long," Warner said graciously. "First, I've got to digress a moment to speak about a personal aspect of this." Then Warner began a lengthy digression that was chock-full of indignation over life's unfair circumstances and the unending misbehaviors of the popular press. "For many years I was married to a wonderful person who is still a very dear and valued friend, and she does have, my former wife, this is twenty-five years ago, twenty-six years ago . . . "—he was taking refuge now in the role of a gentle elderly man looking back on some sadly regrettable life events, perhaps engaging in some nostalgic pentimenti—". . . when we unfortunately parted ways, but she does have a home on the Cape. I was actually married there, to this wonderful woman . . . and every time I try and get into this fray, they conjure up this woman who is very private and three children of mine who occasionally visit the house, and I just think it's fine for anyone to heap criticism on me, but I really resent . . . "—he is becoming the honorable gentleman protecting the honor of a lady—". . . and it's not your organization . . . the local press has seized on this as why I've questioned this thing . . . ."

"But let's put that to one side," the gallant senator said, bringing himself up short, pulling himself out of his reverie, returning to his professional role as a national legislative leader. Then Warner repeated many of the same things the Barnstable town councilors had said so very many years ago. He brought up birds. He brought up fish. He chastised the "environmental community" for "not focusing on this," and, of course, he called Nantucket Sound "a great treasure . . . this magnificent piece of America."

Warner's main concern, he said, was that the Army Corps of Engineers did not have the legislative authority to oversee the permitting of Cape Wind.

Duffy returned Warner's courtliness, saying that he and Cape Wind executives "respectfully" disagreed. In any case, Duffy continued, Washington legislators were addressing that very matter even as Warner spoke. The National Energy Policy Act was finally, after several years of bitter fighting, moving forward. When it did, Duffy said, it was likely that the problem of oversight would be solved. At the time of Duffy's discussion, it appeared as though the Minerals Management Service, under the Department of the Interior, would receive legislative authorization to oversee Cape Wind as well as other offshore wind projects.

Moreover, Duffy explained, the company had prevailed in federal court several times on that very issue of authority under current law.

"Well, we'll see what comes out of that," Warner said, a barely perceptible skepticism creeping into his gentle cadences. "We'll continue to work away . . . ."

To some who heard it, that last sentence seemed to contain an undercurrent of foreboding. Days earlier Warner and his senate colleague Lamar Alexander, junior Republican senator from Tennessee, had introduced a bill on the Senate floor with the Orwellian-Newspeak title of the "Environmentally Responsible Wind Power Act of 2005." This act would deny federal subsidies to any wind turbines built within twenty miles of an American shoreline, within twenty miles of any national park, national military park, national wildlife refuge, or World Heritage site. If a wind project were proposed within twenty miles of a neighboring state's border, the bill would give that neighboring state the simple right to veto the project if that neighboring state didn't want to look at the turbines.

Alexander had delivered an impassioned speech on the Senate floor, in which he said that wide-scale construction of wind turbines would result in "the wholesale destruction of the American landscape," a claim that sounded somewhat Chicken Little–like, even to people who were not necessarily wind-power supporters. "These wind turbines are not your grandmother's windmills, gently pumping water from the farm well," Alexander warned on his Web site.

Republican Senator Pete Domenici, then-chair of the Senate Committee on Energy and Natural Resources, at one point handed Alexander a miniature wind turbine. The "Environmentally Responsible" act was referred to committee, a locale from which it was unlikely ever to emerge.

What was so magical about the number "twenty?" wags wondered. If Lamar Alexander had his way, apparently no wind turbines would be built anywhere within the boundaries of the United States of America. The reason for Alexander's hostility remained unclear, although several activists did note that Alexander had done much to support the coal industry. Another explanation brought forward was the fact that Alexander owned a piece of undeveloped land on Nantucket Island, valued at up to $1 million. Observers wondered: Could Alexander's socializing with the Island's gilt-edged residents and listening to their misinformation and complaints about Cape Wind have influenced him to propose such sweeping antiwind legislation?

Meanwhile, throughout the summer of 2005, the passage of the national Energy Policy Act trudged forward. Included in it was the decision mentioned by Cape Wind's Dennis Duffy to John Warner during his congressional testimony in May—the decision that Minerals Management Service would be responsible for overseeing the operation of and collecting lease payments from offshore wind facilities, including Cape Wind. Many controversies held up passage of the Energy Policy Act of 2005, including the matter of drilling in the Arctic National Wildlife Refuge. For years, the Alaskan delegation and many House of Representatives members had wanted to open the refuge to oil and gas drilling. However, the necessary votes to get such legislation through the Senate did not exist, in part because of the opposition of Massachusetts's senior senator Ted Kennedy.

As the battle over drilling in the Arctic continued, the summer itself became hotter and hotter. In New England, electric grid managers were hard-pressed to find the power necessary to operate the region's air conditioners. By summer's end, just before Hurricane Katrina hit the nation's shoreline, the New England Independent System Operator had registered eight "top demand days" over the course of several grueling heat waves. The system kept setting all-time records for electricity consumption. Power watches were declared, then lifted, as grid managers juggled the system to keep it operating. Had Cape Wind been built, it could have been of immense assistance. On some of the hottest summer days, Cape Wind's data tower showed that, because of high wind speeds, the project would have been producing as much as 300 to 400 megawatts, or enough to greatly ease the anxieties of worried ISO executives.

With temperatures in the nineties across New England for several days at a time—a previously somewhat rare but now more and more typical situation—those who could afford air-conditioning were reluctant to turn it down. Those who could not afford the comfort of electricity that was upwards of fifteen cents and heading toward the hitherto unimaginable cost of twenty cents a kilowatt-hour suffered as best they could and visited the seashore if they could get time away.

Down in Washington, the air-conditioning was running full tilt as the government stayed in the city that was once nothing but swamps, sweating and struggling and finally putting together the Energy Policy Act. It did not include Arctic drilling authorization, but it did include authorization of the Minerals Management Service as the offshore wind-energy oversight agency. The bill also included a clause that effectively exempted Cape Wind, which had already progressed so far along the regulatory pipeline under the U.S. Army Corps of Engineers, from having to start all over again at square one. The passage of this bill would become key in the Cape Wind struggle.

The energy act was finally passed in mid-August, allowing the happy congressmen and senators to leave the draining heat of Washington for the draining heat in whatever district they might come from, be it Montana or Florida, Maine or California. Even the permafrost in Alaska was melting. Its evergreen forest was dying. All over the United States, the highborn and the hard-pressed alike were suffering from the weather. And no sooner had the legislators gone home than there came the first warnings of a dire hurricane headed for the city of New Orleans, one of the nation's most impoverished urban communities.

When Congress reconvened, post-Katrina, an odd thing happened that made Jim Gordon somewhat wary, but not at first overly concerned. A congressman he'd never heard of, one Don Young of Alaska and one of the oldest and gruffest of the old silverbacks down in Washington, D.C., chaired a House committee working on a bill that was supposed to reauthorize the operations of the United States Coast Guard, one of the few federal agencies that came out of the Katrina disaster looking like it had its head on straight. The bill was supposed to be passed by October 1, after which the guard's official authorizing legislation would be past due.

At the last minute, in mid-September, on the floor of the House of Representatives, Young pushed through without previous notice what Gordon would later learn was called a "floor manager's amendment." This amendment required that the Coast Guard Commandant issue an opinion on whether any proposed offshore wind-energy project would pose a hazard to navigation. Never publicly debated, the clause—stuck in a section entitled "Miscellaneous"—seemed innocuous enough. In the case of Cape Wind, the Coast Guard already had the responsibility for investigating the project's safety regarding sea navigation. To date, no project opponent had come up with any substantive charges regarding navigation safety, although many boat owners had complained that sailing would be more difficult.

Gordon was concerned, but he had no reason to think of Don Young as an enemy. If anything, given that Young was from Alaska, Gordon thought it unlikely that the congressman held a grudge against Cape Wind, which was, after all, a continent away from the man's home state. If anything, given that Ted Kennedy opposed Cape Wind, it seemed logical that Don Young, a dyed-in-the-wool antienvironmentalist and rabidly pro-oil man who earnestly yearned for oil companies to have a right to drill in Alaska's northernmost national wildlife refuge, would want, would hope for, would just *love*, to see a wind farm built within sight of Ted Kennedy's estate.

Gordon and Duffy recalled that low rumbling from Warner—"We'll continue to work away . . . "—but Young's amendment had been pasted on to the Coast Guard bill in the House of Representatives. They couldn't figure out how John Warner or Ted Kennedy or Lamar Alexander could be behind Young's action. And when the Senate's version of the Coast Guard reauthorization bill passed the Upper Chamber, sponsored by Alaska senator Ted Stevens, no similar amendment had appeared.

Jim Gordon, however, was not a man to ignore a shot across his bow. Just in case some mischief was afoot, several New England organizations supporting Cape Wind sent along letters to Congress. The Maritime Trades Council called Young's amendment "bad public policy." The Conservation Law Foundation said the amendment would place an "inexplicable" obstacle in the way of offshore wind development.

Cape Wind supporters hunkered down to wait. As is fairly common, the House and the Senate passed two different versions of the Coast Guard authorization bill, one with the offshore wind amendment and one without. A conference committee consisting of both senators and representatives was convened to reconcile the two versions, again a common occurrence. The joint-chamber conference committee needed to create a compromise bill that would be acceptable to both the House and the Senate. After achieving that compromise, the bill would be reported out of the conference committee and returned to both chambers for approval. At that last stage, approval by both chambers is almost always forthcoming.

Cape Wind executives expected the bill to be reported out of committee quite quickly. The Coast Guard—and the nation—needed the legislation soon. It contained legislation that would authorize the Coast Guard to continue to clean up after the Hurricane Katrina disaster. It would also improve the guard's ability to respond to terrorism threats. Given the urgency of these matters, most observers expected the bill to be brought to both floors any day.

It wasn't. October stretched into November, and nothing happened. The bill remained in limbo, in a kind of unofficial "cold storage," during which time no one could get any solid or clear information as to what was going on. No official word came from the conference committee as to what was holding up the bill's progress. Nor was the conference committee required to make any public statements.

Cape Wind lobbyist Gary Klein of DLA Piper (left) and Cape Wind president Jim Gordon ask Alliance spokesman Ernie Corrigan (center), who crashed a Cape Wind press conference at the National Press Club in Washington, to leave.

Gordon found the situation perplexing. In a democratic country, shouldn't the conference committee be required to make its deliberations at least somewhat public? He asked his Washington lobbyist—Gary Klein of the lobbying firm DLA Piper—to visit Young's office to explain the details of the Cape Wind project and to try to find out what was going on.

Klein got no further than a phone call. Young's staff declined the offer of a meeting. Particularly ominous was the response to Klein's request that Young consider a grandfather clause excusing Cape Wind from the additional regulatory duty on the grounds that the company had already proceeded quite a distance along the regulatory pipeline. "No grandfather clause," was the answer Klein received.

Klein called Gordon. "He's not done with you," Klein warned.

Now Gordon was alarmed. The experience reminded him of Ted Kennedy, years earlier, refusing to meet.

Days turned into weeks, and nothing much happened. For those following the Cape Wind saga life became almost boring—although a great deal of merriment was had when, back on Cape Cod, the Alliance to Protect Nantucket Sound held its annual board-meeting-and-changing-of-the-guard in October.

Things hadn't gone too well for the Alliance in the preceding months. Greenpeace, that lover of knavery and good old-fashioned fun, had appeared on Cape Cod beaches—on Craigville Beach, even—carrying signs in support of Cape Wind and talking up the wind farm to sunbathers. Even worse, joining forces with the local project-support group Clean Power Now, Greenpeace activists had actually publicly *teased* Bobby Kennedy Jr. as he sailed Nantucket Sound in an Alliance-funded boat and railed against Cape Wind for the benefit of newspapers.

There was press, but it wasn't necessarily favorable to RFK Jr. In fact, what was supposed to have been good local press for the Alliance instead turned into a coup for Greenpeace. *Cape Cod Times* editor Cliff Schechtman had left the paper in midsummer and the Alliance just couldn't get any more good news stories. Even the youngest and most naïve reporters had become skeptical where the Alliance was concerned. NBC News carried a short spot showing the Kennedy-Greenpeace hoopla. Locally, support for Cape Wind seemed to be mounting.

Onto this stage swaggered the inimitable Bill Koch—Oyster Harbors property owner, Palm Beach property owner, America's Cup victor (albeit many years previously), major Alliance funder, and yet another of the world's most fabulously rich men. Koch had made headlines earlier in the year up in Boston, when he somehow got clearance to place his America's Cup–winning (albeit rather old) sailboat on the lawn of Boston's much-loved Museum of Fine Arts (MFA). Inside the MFA was a "show" of Koch-owned paraphernalia: paintings and wine bottles and horse's saddles and bridles and Mexican spurs and really cruel things like a bit capable of breaking a horse's jaw, used by men lacking true equestrian skill who must instead rule the animal with force.

"Things I Love" was Koch's chosen title used for the "art" show. The title caught the attention of Boston journalists, who wondered why he loved all those nasty old spurs and horse bits, and who asked about how such "art" could have gotten into the revered MFA.

Having become famous up in Boston—billboards advertising "Things I Love" featuring a famous painting of a nude woman lying seductively on her side were placed prominently alongside the city's major highways—Koch next offered himself as the man who could pull the Alliance out of its doldrums. After all, if he lay siege to the MFA, he could certainly take command of the seaside civil war. Koch became the organization's head, a duty shared with one Christy Mihos, a man who would be caught welshing on some yachting taxes and who would eventually put himself up for governor as an independent candidate.

New Alliance head Bill Koch took over the lawns and halls of Boston's revered Museum of Fine Arts with "Things I Love," an "art" exhibit that included a mishmash of items ranging from rusty cowboy spurs to several of his aging yachts.

After taking over the reins of the Alliance, Koch apparently intended to apply bits and spurs in a headlong cavalry charge, a fact that many gleaned after reading a wonderful interview Koch gave to young *Cape Cod Times* journalist Kevin Dennehy.

Koch reportedly said that in the past, the Alliance had been "goofy," a succinct analysis which Dennehy, in the post-Schechtman newsroom, had the good sense to put in his story's lead. Koch also touted the similarities between wind turbines and exotic dancers. Both, he said, are "seductive." Wind power looks nice at first, Koch reportedly said, but so does the gaze of an exotic dancer. Realizing that readers might be confused by the comparison, the interviewee went on to explain himself: If you yield to temptation, who knows what will happen? ". . . go over there you can get in a lot of trouble, and get a lot of diseases."

Upon reading this, many people snickered. They fastened their seat belts. The upcoming legislative year promised to be a wild ride, a roller coaster of wicked intrigue and great quotes.

There was one serious note that brought a bit of reality to the fun, however. In mid-November, grid managers issued yet another official warning about tight electricity supplies, should another cold snap occur.

Late one Saturday morning in early December, Cape Wind communications officer Mark Rodgers was putting grocery bags into the trunk of his car when his cell phone rang. Rodgers had his toddler in tow and was looking forward to a weekend of domesticity. This hope was immediately dashed when he heard Jim Gordon's voice. His tone was quite serious.

Information had surfaced about the secretive dealings in the conference committee. The purpose behind Young's "floor manager's amendment" had been revealed. It was actually a "placeholder amendment"—an amendment that kept a space open in the bill, but could be substantively changed behind the closed doors of a conference committee. This, in fact, was exactly what had happened.

Things didn't look so good for Cape Wind, Gordon told Rodgers. Gordon had been on the phone for much of the previous evening. His Washington lobbyists had informed him of a rumor that Young had "reworded" his amendment rather considerably. It bore little resemblance to the first version that had been voted on by the full House of Representatives—something that is, astonishingly, permitted under arcane House rules.

Young now wanted to "prohibit the establishment of any offshore wind energy facility within 1.5 nautical miles of a shipping channel or commonly used route for a . . . ferry system." Since Young was the chair of the House side of the conference committee, he possessed a considerable level of power. This new amendment would not be publicly debated, or even, barring some miracle, publicly discussed. In fact, the only reason the language surfaced at all was because an indignant source involved with the work of the conference committee leaked it.

If the bill emerged from conference committee containing this newest version of the amendment, Gordon's lobbyists told him, Cape Wind was pretty much done for—as in, kaput, over, blown away. After abiding by the 1.5 nautical mile limitation, there would be room for only a few wind turbines on the shoal, which would make the project, in Gordon's opinion, financially nonviable.

In a moment of panic, Rodgers wondered if he would soon need to look for another job. Then he resolved to fight. He drove his toddler home and let his wife know she wouldn't be seeing him much in the coming days. He and coworker Rachel Pachter began calling everyone they could think of, people who had maritime expertise and who would be willing to fire off an e-mail letter to members of the committee. Rodgers and Pachter were gratified to find that there were plenty of such people, people who hadn't necessarily signed on as active project supporters, but who were offended by Young's backroom tactics and by the Alliance's claim that the wind turbines would be dangerous to ships. It just wasn't so, the letter writers explained, adding that there existed no evidence that the turbines posed a hazard to shipping, that in Europe offshore turbines were often quite close to shipping lanes, that oil rigs in the Gulf of Mexico were allowed to be within 500 feet of a shipping lane, and so on.

For his part, Gordon himself had a choice to make as company president. As a lobbyist, Klein, almost as indignant over the reworded amendment as Gordon, had explained as much as he knew about what was going on behind closed doors in Washington. Listening to him, Gordon felt as though the air was sucked out of the room, as though the walls were closing in. How could some congressman from Alaska, some guy named Young, heading up an obscure committee on the Coast Guard, have the power to suddenly destroy, in one fell swoop, five years worth of work?

First, Gordon thought to himself, opponents on Cape Cod had come after him. Then it had come from Boston, from the Romney administration.

Then, from Virginia. Now, here was some guy coming out of the Alaska tundra with a Bowie knife between his teeth.

Didn't the passage of the Energy Policy Act only a few months earlier, which authorized the Minerals Management Service to oversee offshore wind siting, have precedence over something pushed through by some guy who wasn't even on the energy committee? Klein's answer had been short: "Not necessarily."

"Unbelievable," said Gordon. Scenes kept going through his head, of all the past public hearings, of his attempts to talk with editors at the *Cape Cod Times,* of the gathering support of groups nationwide who wanted to see his project move forward. He had a teenaged son, a toddler at home, and another baby on the way. He was middle-aged now. Five years had passed. Wouldn't it just be easier to devote his time to his wife and children, to traveling to the places he'd always wanted to see, to having dinner with friends? "Maybe this has just been a fool's errand," he thought to himself.

The ways of Washington are wond'rous indeed. And among the most wonderful is the ability of some longtimers to amass power that would, if converted into electricity, set the world aglow for years into the future. Don Young was such a man, or at least he would have others so believe. A California native, Young moved to Alaska in 1959, just after it gained statehood status. He sought the adventurous life of a backwoodsman, riding on tugboats and looking for gold and hunting big game. Some say that having not found such a life in Alaska, the backwoodsman wannabe came to Washington. Young was elected to Congress in March 1973, mostly on the basis of his promise to ram through legislation that would make possible the controversial Trans-Alaska Pipeline.

When the Cape Wind controversy hit town, Young was the third most senior Republican in the House of Representatives; the eighth most senior member, if you count Democrats. He is a proud hunter who displays many trophies on his Washington office walls. He is an earmark-driven man. He generally has a short, white beard, but sometimes allows the beard hairs to scraggle out a bit, possibly just to remind people that his beard is not the collegiate or academic sort. He is not particularly liked by people outside of Alaska, and his first, second, and third legislative priorities are drill, drill, drill in Alaska, Alaska, Alaska.

Young's philosophy of government is based somewhat on the doings of Lyndon Johnson and Bobby Baker. Fondly calling himself "Congressman for All Alaska" (there are only enough people in Alaska to warrant one congressman), Young was the chair of the House Committee on Transportation and Infrastructure—an immensely lucrative committee, in that all kinds of national corporations specializing in highway construction and other kinds of heavy contracting are anxious to contribute as much as possible to a chairman's campaign coffers. The committee deals with an immense variety of subjects, ranging from highway construction to aviation to maritime issues. As chairman, Young had the right to oversee anything that concerns the Coast Guard, which has its own subcommittee, the Coast Guard and Marine Transportation Committee.

Nevertheless, Young's interest in matters relating to a small body of water a continent away seemed curious. The man himself was a model of sphinxlike behavior, refusing to discuss the issue. Indeed, when reporters first called his office to ask about Young's Cape Wind mischief, the congressman's staff denied their man was up to anything. Cape Wind, they said, was "paranoid."

This turned out to be a lie. Theories explaining Young's motivation were plentiful. Some claimed he had made a deal with Kennedy, who had agreed to give in to Young's desire for drilling in the Arctic National Wildlife Refuge. Others pooh-poohed that train of thought, although sadly, given Kennedy's highly emotional opposition to Cape Wind, this theory was not altogether beyond the realm of possibility. Others suggested that Young might be doing a favor for Billy Lee Evans of Kessler and Associates, a former congressman and now a lobbyist for Bill Koch's energy company through its association with U.S. Strategies. That lobbying firm, used by Koch, sometimes hosted Ted Kennedy on its private jet, allowing the lobbyists one-on-one time consistently denied Jim Gordon.

Others figured it boiled down to Young's antipathy for Theodore Roosevelt IV, Lehman Brothers managing director. Following the radio show in which Roosevelt and Gordon had countered Corrigan's anti-wind arguments, the financier and the energy entrepreneur had continued their discussions. More than a year later, when Gordon began soliciting investment firms, Lehman Brothers was among those that responded. Cape Wind Associates chose Lehman Brothers because Roosevelt seemed to have a deep understanding of what the project could achieve. Roosevelt had devoted much time and effort to climate-change issues. He seemed to Gordon a good choice to lead Cape Wind's project financing. Additionally, Lehman Brothers itself already had a demonstrated record of putting together successful debt-and-equity packages for energy projects.

Bill Koch's U.S. Strategies lobbyist Jeff Peterson (left) debates Cape Wind with Cape Wind lobbyist Gerry Harrington of Capitol City Group.

Don Young was no fan of Ted Roosevelt. When a reporter in Alaska asked the congressman why he had gone after the Massachusetts energy project, Young responded: "Have you seen who is financing it?" Roosevelt, a good-government supporter like his great-grandfather, had made several choice public comments about Young over the years, including a speech to a Republican gathering in which he implied that the party should be ashamed to call Don Young a member.

As Gordon mulled over his options after speaking to his Washington lobbyist—should he continue to fight for Cape Wind, or should he move on to something else—he thought about Don Young. Was it worthwhile going up against the Alaska congressman? Did he want to spend the company's money on a fight that, advisers warned, probably could not be won? What were the percentages? What was the point?

Conversely, Gordon realized that he was no longer alone in this fight. Over the past several years, more and more supporters had come to his aid. Union members, clean-air advocates, good-government advocates, and many other groups were increasingly aware of the Cape Wind project. Initially, many national environmental groups had stayed at arm's length because they were uncertain about the potential impacts of the project. Now, though, after having read the Environmental Impact Statement, many of these groups had decided that they could support the project, if certain scientific parameters were met, including extensive follow-up studies regarding birds, marine mammals, and other sea life. Now those groups said they were committed to Cape Wind moving forward. Did they mean that? Would they put in the 110 percent necessary to get through the process? Gordon hoped they would.

It was a long evening for Cape Wind's president, but by the time he called Rodgers, Gordon had made his decision. He would go for the sunshine, opt for the high road, take his case to the nation's journalists and make sure everyone knew what was going on. If he was going to go down, his demise would be public.

What resulted from this decision was the incredible, implausible, perhaps even unprecedented formation of a team of odd bedfellows that included participants from all walks of Washington political society. Greenpeace found itself working with General Electric. *Washington Post* editorial writers found themselves agreeing with *Washington Times* editorial writers. Liberal Republican senator Lincoln Chafee of Rhode Island, an outspoken leader when it came to renewable energy, found himself aligned with conservative Republican congressman Joe Barton of Texas, a man who was certainly no fan of the Clean Air Act and who would probably prefer to die of thirst in a Texas desert than to escape by driving a hybrid to a water hole.

The ensuing six-month effort to keep Cape Wind from being quietly rubbed out was centrist and completely bipartisan. If there wasn't a sense of optimism about the outcome, there was certainly a sense of enthusiasm. Across the spectrum of participants, indignation regarding not just the amendment language but the Machiavellian manner in which the amendment language was constantly changed and in which the democratic process was subverted, fueled the energies of even the most jaded Washington observers.

And perhaps most remarkable of all—the whole effort concerned one relatively small energy project, which, if successful, would result in a significant amount of economic gain for a small group of investors. Members of the coalition, however, also saw in the success of Cape Wind various other gains, depending on their various points of view. The environmental groups saw both a significant step forward in the problem of climate change and, for southeastern Massachusetts, a significant step forward in cleaning up air and water pollution. General Electric, of course, saw the project as a significant step forward for the wind industry, in which it had heavily invested through the acquisition of Enron Wind after Enron imploded. For both *Post* and *Times* editorial writers, the issue had become a "good-government" battle. Even some of Cape Wind's lobbyists, offended on a visceral level by the attempt to kill a project that could offer some genuinely important changes in the nation's energy infrastructure, were involved on an emotional as well as a professional level.

For the many politicians who gradually became involved, the battle was about many different issues. Quite a few climbed onboard the Cape Wind bandwagon because they sincerely believed in the good-government debate, particularly prevalent in Washington just then because of lobbying scandals centering around the infamous Jack Abramoff. Others wanted to change the nation's long-stagnant energy infrastructure. For some, the driving issue involved a turf war. After years of debate the Senate and House energy committees had crafted an energy policy bill, only to have it enfeebled via guerrilla warfare tactics. What was the point, they wondered, of having worked so hard?

Step one for the Cape Wind team was outing the perpetrators. The rumor mill had revealed Young's strategy, but proof was lacking and Young's staff was lying. Under these circumstances, the press was somewhat limited in what it could write about. Grassroots groups, however, could get involved. A wide range of advocacy organizations sent blast e-mails to their members, asking them to contact Washington legislators. These groups focused particularly on the seven senatorial members of the conference committee, because Cape Wind strategists assumed that under Don Young's heavy-handed leadership, members of the House of Representatives would obey their committee chairman.

In the lead of this coalition of advocacy groups was the Cape-based group Clean Power Now, accompanied by the fun-loving folks at Greenpeace. For several years, Greenpeace had made Cape Wind a priority. Now the organization decided to focus on the project even more. The organization's USA office head, John Passacantando, was a smart-as-a-whip guy compelled by a level of righteousness not often found in the halls of government. Passacantando, a lithe outdoorsman with an agile build and a slight air of intellectual fervor, had moved over from Wall Street to the environmental movement, bringing with him plenty of useful street-fighting skills and a high level of competitiveness. Under his leadership, which began in 2000, Greenpeace was no wallflower.

It's not always easy at first to peg Passacantando as an "environmentalist." He holds several degrees in economics and still talks like a Wall Street trader. "I want in. I want in this game," he said, hearing what Young was doing.

Passacantando believed that Greenpeace—which does not receive government grants, does not take corporate donations, claims 5 million members worldwide, and is not beholden to any source of money save what dribbles in from its grassroots members—had a special role to play. The organization didn't worry about stepping on anybody's toes, includ-

Greenpeace USA head John Passacantando led his organization in its active role of ridiculing Massachusetts senator Edward M. Kennedy for his opposition to Cape Wind.

ing those of the Kennedy family. Passacantando intended to play hardball by teasing Ted Kennedy mercilessly, until he laughed, or until he cried. Unlike many other environmental organizations, the renegade Greenpeace had no major funding sources beholden to the Kennedy family. In Passacantando's opinion, Greenpeace, owing no favors, could fight a take-no-prisoners battle.

"Ted Kennedy doesn't get a special pass on this," became the Greenpeace litany. "Global warming is too important."

Throughout January and early February 2006, while grassroots groups called elected officials and visited Washington, journalists waited for the bill to be reported out of conference committee, so they would have something substantive to report. Day after day, the bill was said to be coming. Day after day, it was not reported out. The ability of Washington's elected officials to behave so secretively continued to incense Gordon, so much so that the immediacy of the emotion would remain with him for many months into the future.

Behind the scenes, though, a lot was going on. Gordon and company were on pins and needles, flying back and forth to Washington, plying the halls of Congress. Although they didn't have much to go on in the way of hard evidence or a specific document, Gordon's lobbyist Gary Klein kept them informed. He had been told by many different people that Young was absolutely unbendable on the matter of Cape Wind. Exactly why this was so remained unclear. Congressman Bill Delahunt, a man of rather inconsiderable power in Washington, had been lobbying his congressional colleagues to kill the project. Ted Kennedy had also been approaching his colleagues, making personal appeals, explaining that the protection of Nantucket Sound was important to him and to his family, and that his family deserved this, given the many sacrifices the family had made for the nation. In Washington politics, friendship counts for a lot. While Delahunt could be somewhat placated but mostly ignored, it wasn't easy for many old-timers, who had known Kennedy for decades, to turn away from a man who was an old friend.

Cape Wind decided to counter those personal appeals with a vocal group speaking out for principle, rather than friendship. America needed to change, these people said. That the change could—*should*, even—start with Cape Wind. Amid the din of public uproar, many of the younger, newer elected officials tried to listen to what Cape Wind executives had to say.

Among those was Maine Republican senator Olympia Snowe. Elected to the Senate in 1994, Snowe was not a member of the bipartisan old-bull network that seemed to be running the antiwind movement. Born just after World War II, she was a baby boomer, a natural born politician who had the odd distinction of being the first woman ever to have served in both the upper and the lower houses of the Maine legislature and then, in Congress. She had "inherited" her seat in the Maine House of Representatives in 1973 when her husband, then a representative, suddenly died. Only twenty-five at the time, she has held public office ever since.

Snowe became a key member of the conference committee. Four of the committee's seven senators—Ted Stevens of Alaska, Trent Lott of Mississippi, Gordon Smith of Oregon, Dan Inouye of Hawaii, Maria Cantwell of Washington State, Frank Lautenberg of New Jersey, and Olympia Snowe—needed to sign the bill before it could emerge from conference committee and go to the floor. Stevens, Lott, and Inouye, a Democrat and the third most senior member of the Senate, were considered hopeless cases, in that they were likely to do what Ted Kennedy wanted. Cape Wind executives and supporters pinned their hopes on Smith, Cantwell, Lautenberg, and Snowe. If Snowe refused to sign the bill with its anti–Cape Wind amendment, strategists hoped, the other three might follow her lead.

Snowe seemed a good choice. As a Republican, it would not harm her to gain a reputation for opposing Ted Kennedy. Moreover, Snowe—and the journalists in her state—were already familiar with the Cape Wind battle. Strategists opted for a three-pronged approach. Rhode Island Republican Lincoln Chafee appealed to Snowe as a colleague, asking her to hold steady and to refuse to sign the bill. Maine voters on the e-mail blast lists of the grassroots groups supporting Cape Wind flooded Snowe's office with phone calls. Maine's newspaper editorial writers called on Snowe to show leadership and hold the line against the Cape Wind ambush. "Snowe ought to vote to reject Young's amendment and let Cape Wind's alternative energy project succeed or fail on its merits," wrote the *Portland Press Herald* at the end of February. "If Snowe and the others do not succeed in removing Young's language from the final bill, they will share blame for killing one of the country's most important renewable energy initiatives," wrote the *Boston Globe.*

Seeing Snowe's predicament, other senators also hesitated to sign. Days turned into weeks, and Young could not move his bill out of the conference committee. Each week of delay was a small victory for Cape Wind, since the more people were aware of the situation, the less power the opposition had. Sunlight is a great disinfectant. Grassroots indignation, perhaps fueled in part by lobbying scandals and spiking gasoline prices, mushroomed. Elected officials who would have supported Kennedy quietly were reluctant to do so while the issue remained so much in the public eye.

The Cape Wind team continued, however, to suffer from a serious handicap: the secrecy of the conference committee. They knew something about what Young was up to, but they lacked the smoking gun—the definitive proof that Young was out to ambush Cape Wind specifically. Without this evidence, the battle was fought in a vague netherworld and was frustratingly limited to he-said/she-said charges, little of which made for good newspaper copy.

But then Young made a decisive error in strategy. Frustrated by his inability to get his bill released from conference-committee purgatory, Young distributed a five-page dear-colleague letter purportedly explaining why the amendment was necessary. Although it claimed that he was interested in the matter of navigation safety throughout the nation's waterways, the letter documented Young's targeted animus for Cape Wind. Of the nineteen paragraphs in the letter, six paragraphs focused entirely on only the Nantucket Sound project. Moreover, Alliance sound bites featured prominently in the document. He claimed, for example, that the state protected Nantucket Sound as a "marine sanctuary."

With this document as Exhibit A, the Cape Wind team now had something they could hang their hats on. A firestorm ensued. Cape Wind immediately wrote a reply letter, correcting Young's misinformation. They began another round of contacting journalists and patiently explained the facts of the project. News stories appeared across the continent and even in foreign publications. Journalists began asking why Young would restrict the future of offshore wind in America so severely, when it was readily apparent that he only wanted, for some inexplicable reason, to block Cape Wind.

Others began writing to Washington. Matt Patrick wrote complaining that the amendment "would implement the most restrictive program in the world to discourage offshore wind energy development." Deval Patrick, a grassroots Democratic candidate for Massachusetts governor who had had the courage to support the wind farm despite Kennedy's hold on the Massachusetts's Democratic Party, called the state's three top elected officials—Romney, Lieutenant Governor Kerry Healey (his upcoming gubernatorial opponent), and Attorney General Tom Reilly—the "gang-of-three" for opposing Cape Wind and not speaking out about the Young amendment. Massachusetts state legislators from both the House and the Senate wrote to Washington asking why national officials were meddling in a project that the state's elected officials had worked so hard to move forward.

Massachusetts gubernatorial candidate Deval Patrick begins building his grassroots campaign by signing on in support of Cape Wind, standing below the Hull I turbine.

It was as though, for one brief moment, 1960s-style activism returned to the nation's capital city. Union members besieged the Washington offices of their elected representatives. Project supporters began talking seriously about a march on Washington. One businessman offered to finance an advertising blimp that would float over the Capitol Dome. John MacLeod from the town of Hull wrote to complain that Hull, which was about to install a second wind turbine, intended to become totally powered by wind energy and that Young would make that goal impossible to achieve. Lincoln Chafee wrote to colleagues that the Young amendment would be "devastating" to his state's renewable-energy goals. True Majority, a good-government group started by Ben Cohen, one of the founders of Ben & Jerry's Ice

Cream, notified its 60,000 members to take up arms by dialing their Washington legislators. The *New York Times* wrote "Sneak Attack on Cape Wind." The *Boston Globe* wrote "Capitol Hill's Ill Wind." Alaska papers ran stories about Young's meddling.

The onslaught went on and on, letters and phone calls and news stories filled with outrage. But perhaps most remarkable were documents emanating from two particular individuals. Heavily pressured behind the scenes by boards of directors from several national environmental groups, Robert F. Kennedy Jr. sent a letter on his own letterhead that said: "It is simply not good government to insert such a significant last-minute amendment" in a conference committee bill. And the junior

The town of Hull's new 1.8-megawatt wind turbine translates the energy of the wind into energy to run the town's lights at night. Most residents say they like the turbine, which is easily visible from many backyards.

senator from Massachusetts, John Kerry, issued a very short press release decrying the strategy on the basis of good-government principles. Kerry made sure, however, not to word the release in a way that might be perceived as outright support for Cape Wind.

Gordon took heart from this outpouring of anger at Young. "What doesn't kill you, makes you stronger," he said. "We'll survive this assassination attempt. They've done us a favor because they've crossed the line. They've elevated this now into a national issue."

His schedule was grueling. He'd been flying back and forth to Washington each week, spending his life on the phone, rallying his troops while trying to find time to be with his wife, planning for the birth of the couple's next child. Every week rumors spread through Washington and beyond that the bill would be reported out. It wasn't.

The day the *New York Times* editorial ran chastising Young and calling upon Senator Snowe to refuse to sign the bill, Gordon was elated. He had talked to the *Times* editorial writer for quite a long time, had pleaded, begged for help. "Please, sir," he implored the editorial writer, "we need your help."

When it ran, Gordon experienced a kind of low-key euphoria. The editorial was much more supportive than he had dared hope for. It seemed to him, a kid whose father owned a corner store in a lower-middle-class section of Boston, a profound example of the value of American democracy that one of the nation's largest and most influential newspapers had taken up his standard.

But later that same late-February day, the Washington rumor mill began feeding information to Cape Wind supporters that Young had changed the amendment's language yet again. Rather than wording the amendment to affect the offshore wind-energy industry as a whole, Young had specified that the amendment would be directed only at

Cape Wind. The strategy was apparently designed to lessen the blowback from Young's earlier proposal. Another smaller project of roughly forty turbines was in the pipeline for the southern shoreline of Long Island in New York State, and elsewhere in the country various groups were working on offshore wind proposals. This new, very targeted language might help lessen the concerns of interested parties outside of Massachusetts.

Young's newest idea, specifically naming the Nantucket Sound project, wasn't meeting with too much enthusiasm in Washington, Gordon was hearing. But who knew? Congressmen are famous for making a big deal about keeping out of the other fellow's territory. Maybe they would decide it was just easier to let Kennedy and Delahunt have their way and kill the project behind closed doors, where elected officials wouldn't, they hoped, be held to public account.

It was a roller-coaster day for Gordon, but it wasn't over yet. That evening he walked through the plate glass doors of New England Cable News all by his lonesome self. It was 8 o'clock on a night that was very

Jim Gordon and Cape Wind opponent Cliff Carroll discuss opponents' lobbying activities in Washington, D.C., with New England Cable News host Jim Braude.

dark. It was bone-chilling damp and the sky was spitting snow. Gordon looked completely burnt out, run through the wringer many times over. Those who saw Gordon's drawn face believed he would quit, not realizing that he possessed a stubbornness that was nearly pathological.

The previous day, a producer for the cable station's talk-show host Jim Braude had asked Gordon to do the show. By then, it wasn't first on his to-do list. That would be a good night's sleep. But he was stoic about it: "When they call—you go." By the time he'd walked through those plate glass doors, he'd spent yet another day glued to the telephone, then had had a business dinner with executives from another company.

"Where's your entourage?" asked a reporter, who wondered why Gordon was coming to the television station by himself.

"Entourage?" Gordon answered, surprised by the suggestion that he might have a companion along. "I don't have an entourage. . . . Everyone else is at home asleep."

Gordon was a juggler, addicted to keeping all the balls in the air for as long as possible, but over the past months, the balls had become iron weights.

The next day in Washington, the picture changed yet again. Gordon's hopes were rewarded. Two of the Senate's leading voices issued a warning shot across the bow of the Alaska legislators. A letter issued jointly by the senators from New Mexico, Pete Domenici and Jeff Bingamen, Republican and Democrat, respectively, and chairman and ranking member, respectively, of the Senate Committee on Energy and Natural Resources, said the Young amendment conflicted with the energy bill passed last year by Congress. The letter, immediately released to the public, was mildly worded but made clear to the Alaska delegation that the leaders of the energy committee felt . . . interfered with.

The senators' entry into the battle signified a major turning point in the congressional clash. It provided journalists with yet another news hook for yet another round of news stories. It spurred on grassroots activist groups to continue their efforts. It caught the attention of other senators and congressmen, who realized that a showdown between power centers seemed to be developing.

The state of Alaska was not faring well in the national press. The outcry against its legislators had become so intense that its Republican governor, Frank Murkowski, issued a press release in mid-March saying that he'd hired a marketing company to do a national advertising campaign that would address the "sorely distorted" view other Americans were developing of the state. Americans had begun to develop "negative perceptions" of Alaska, the governor worried. Among other issues, Young's reputation for "earmarking" large sums of money for his pet projects in Alaska was becoming an embarrassment.

Those negative perceptions were about to worsen, as the Alaska delegation dug in its heels regarding Cape Wind. At the end of March and the beginning of April, while the conference committee continued to argue over the Cape Wind controversy, Alaska Republican senator Ted Stevens, then head of the Senate committee overseeing the bill's progress through the legislative process, stepped in and took over from Young. No one knows exactly why Stevens did this, but it could simply have come down to the fact that Young, already in trouble for earmarking monies slated for his pet projects, no longer wanted to be in the spotlight. At one point, a Young aide had told a reporter questioning Young's interest in the Cape Wind matter to call Massachusetts representative Bill Delahunt. Perhaps Young just wanted to wash his hands of the matter.

When Stevens stepped in, he proposed yet another version of amendment language. Stevens's version gave Massachusetts governor Mitt Romney the right to simply veto Cape Wind. Stevens's new "compromise" language said: "An offshore wind energy facility may not be constructed in the area commonly known as Nantucket Sound . . ." if "the governor of an adjacent state" opposes the project. Romney soon made clear that he would make use of that veto power, if the language were passed. The first version of this amendment said the decision to veto could not be made in an "arbitrary and capricious" manner. But over a weekend, the language became even bolder. No reason for the veto needed to be provided at all.

It appeared as though Kennedy, Romney, Stevens, Young, and the rest of the anti–Cape Wind crowd had decided to solve their problem by ignoring the rule of law. It was a magnificently brazen play, a bold end run around the democratic process of such imagination, such assumption of royal prerogative, that many observers were, at first . . . speechless.

All pretense vanished. Young had said he was merely concerned about maritime safety matters around the nation, but now the average person could easily understand that the goal really was to kill Cape Wind.

The heads of the Senate energy committee, Domenici and Bingamen, continued to find the intrusion into their territory offensive. "It would be folly for us in Congress to talk about breaking our addiction to foreign oil and, at the same time, pass laws that stymie our own production of clean and renewable energies here at home," Domenici declared in a written statement. Bingaman called for a "clean bill" for the Coast Guard, writing, "If a special-interest provision to veto a single project by earmark in a conference report succeeded, it would make a mockery not only of all the statements in Congress about the need to strengthen America's domestic energy security, but also our statements advocating lobbying reform in Washington."

Stevens and the others tried to deflect the mounting anger. They focused on getting the bill out of conference committee, regardless of the building backlash. The conference committee senators who had refused to sign were pressured into conforming. Trading went on in the backrooms of the conference committee—nobody knows what was given to whom—but, as if in a deluge, the dissidents caved.

Finally, on April 6, Gordon Smith agreed to sign. He was followed by Maria Cantwell, who was said to want some ships that were provided in the bill. Finally, at last, even Snowe capitulated. With that, the

Coast Guard bill with the Stevens amendment passed out of conference committee.

Once a bill comes out of conference committee, Gordon's Washington lobbyists warned, the legislation can almost never be changed. Because the bill had originated in the House, it would, by custom, first go to the House for a final vote. Floor debate was unlikely, and the vote would most likely happen quite quickly. By the time the bill got to the Senate, it was unlikely that officials in that body would care enough about Cape Wind to take up debate time to straighten out the problem. To Gordon, the most important thing in the bill was the future of Cape Wind, but to elected officials and the nation at large, the bill's importance focused on the Coast Guard. In the larger scheme of things, Cape Wind was just a sideshow.

Cape Wind was told to prepare for the worst. Passage looked inevitable. Even the chairman of the Senate energy committee, Pete Domenici, figured the battle was over. "It's too bad," Domenici told a reporter. "I don't think we can do anything about it." For Gordon, it was the lowest point in his five-year battle.

The day after the news broke, on a gray, drizzly, early-spring evening, Jim Gordon entered the living room of a Nantucket house where dejected supporters were gathered. "This is the way revolutions start," he said. He thrust his curled right fist over his head. Dancing on the balls of his feet, he looked like a cross between Rocky Balboa and Che Guevara.

Despite the seeming disaster in Washington, he was buzzed. An Associated Press reporter had just called him to say that Ted Kennedy had finally admitted publicly to being the power behind the Young-Stevens amendment. Gordon told jokes. "I should get the award for the worst businessman in America," he said. "Five years into this project, and we've spent more than $20 million and we still haven't made a penny."

He laughed. "I got the *Washington Post* and the *Washington Times* to agree maybe for the first time ever," he told supporters, referring to the editorials in both papers that supported Cape Wind's right to fair process. "Maybe I should go to the Middle East."

He coaxed his supporters to lighten up a bit. He brought up Bill Koch's comparison between wind turbines and strippers, a reference that never failed to amuse those following the saga. "This is not the end of the story," he said. "It stinks. It's rotten. This has all the elements of what's wrong down in Washington." He talked for an hour. "I know I've spoken too long," he said. "I apologize. I'm working on adrenalin."

But he couldn't stop. "Grassroots is what it's all about. Grassroots effort can beat money and power . . . ." By the time he was done, his supporters had forgotten about the dire situation in Washington and focused instead on Ted Kennedy's public outing. They were ready for Phase Two.

CHAPTER FIFTEEN

The Tide Turns

Democracy is not a spectator sport.

Motto on Senator John W. Warner's Necktie

When the bill passed out of conference committee, it needed to be filed in the House of Representatives before it could come to a vote. That housekeeping task belonged to Alaskan Don Young as the head of the committee.

Gary Klein couldn't believe his good luck.

Young did not move forward immediately, but took several days to file. Confident as he was of his own power, he must have believed the brief delay wouldn't matter. When he did take action, it was on the eve of Congress's Easter recess.

Once again, Cape Wind opponents hadn't counted on the determination of Jim Gordon and the level of commitment of the pro–Cape Wind forces. Filing just before the recess provided Cape Wind strategists a crucial window of opportunity. They used the delay caused by Young himself and the extra recess days to regroup their forces.

While elected officials took a vacation, project supporters doubled their work commitments. The team had been having daily 3:30 conference calls, during which all participants—union representatives, grassroots environmental activists, lobbyists, Cape Wind staff, good-government activists, and various interested individuals like Ted Roosevelt—brought each other up-to-date on the day's activities.

After the Coast Guard bill was reported out of committee, the team continued those calls. The first order of discussion was the long odds they were facing. Everyone understood that they had little chance of succeeding. No one wanted to call it quits.

The second order of business was to cobble together a new strategy. A major part of that strategy was to come from a somewhat unanticipated supporter. The Cape Wind team had a friend in Washington they hadn't yet met.

Around the time when James Jackson Storrow was starting the Boys Clubs of America, ex-president Theodore Roosevelt was bucking politics as usual in the Republican Party by running, yet again, for president. His banner was the Progressive movement sweeping the nation.

Most Republicans, even those who genuinely liked the Old Lion, were afraid to side with him against the incumbent, Republican President William Howard Taft, who had once been Roosevelt's friend. Robert Perkins Bass, the Republican Governor of New Hampshire from 1911 to 1913, was an exception to that rule. Not a man to abandon a friend, Bass was also in the forefront of the Progressive movement. As a state senator from Peterborough, New Hampshire, he shepherded through the state legislature a bill that instituted the first direct primary in a state east of the Mississippi. He led a state conservation organization that protected the state's White Mountains, today famous worldwide as one of New England's premier outdoor destinations. He supported passage of the first Workmen's Compensation Act. He ran for governor, calling himself a Roosevelt Republican, and won.

Throughout the early months of 2006, the grandson of Robert P. Bass, Charlie Bass, Republican congressman from Peterborough, New Hampshire, had followed the various iterations of the amendment pushed through conference committee. A member of the Edgartown Yacht Club on Martha's Vineyard, a club much favored by Ted Kennedy and his crowd, Bass had listened to the whining about Cape Wind coming from the yachting crowd for years. When the Stevens amendment allowed Romney to veto the project, Bass knew that immediate action was necessary. Otherwise the project would almost certainly go down. He had never met Jim Gordon, but Bass believed that the project itself was important for the region's—and the nation's—energy future.

On April 11, only a few days after the conferees caved, Congressman Bass issued a press release that made clear his intent. It was a kind of gentlemanly declaration of war. "Because the New England electricity grid is interconnected and interdependent, failing to ensure adequate resources for tomorrow's needs could affect my constituents in New Hampshire and force the entire region to pay higher rates if we are forced into using costlier fuels," said the release. "The Coast Guard bill is an improper place to be setting the nation's energy policy."

Reading the release, Cape Wind officials were taken by surprise. "Who is this guy?" asked Mark Rodgers. To Rodgers, with no knowledge of Bass's family legacy, the entry of the New Hampshire Republican seemed like pennies from heaven. Why was he going out of his way to help? "He's a man of conviction. He just does this stuff sometimes," Rodgers was told by lobbyists he asked.

For Cape Wind, Bass's entry may have come out of the blue, but those who knew Charlie Bass knew that the press release was entirely consistent with the congressman's emerging leadership style. Bass chaired the Tuesday Group, a gathering of moderate House Republicans interested in reform. Like his father and grandfather before him, Bass is interested in change. In Peterborough, a wood-pellet boiler heats his home. The southeastern region of New Hampshire, where Peterborough is located, is a hotbed of renewable-energy advocacy. Bass is a neighbor of the publisher of Renewable Energy Access, an international online news service. Bass himself has consistently pushed for cleaner-energy technologies. He opposed drilling in the Arctic National Wildlife Refuge. He supported the development of more efficient solar cells and panels.

Like his grandfather before him, Bass made a habit of stepping out in front of the crowd. In early January, Bass and a colleague circulated a letter calling for then-House Majority Leader Tom DeLay of Texas to step down from his post. Delay, a close associate of disgraced lobbyist Abramoff, faced corruption accusations but had struggled nonetheless to stay in power in Washington. Bass, along with other moderate Republicans, was successful in convincing Delay to relinquish his majority leader post.

Additionally, Bass and Republican senator John McCain led the fight for campaign finance reform. Both were outspokenly committed to slowing the flood of influence money that had been overwhelming Washington government in recent years.

Bass was elected to Congress in 1994. The baby boomer began his political career in 1982 as a state representative and then went on to serve as a state senator. He was defeated for re-election in the Democratic triumph of 2006. For a time, he was the only New England Republican to serve on the House Energy Committee, which, said his staff, led him to think of himself as representing the entire New England region regarding energy matters.

The combination of inherited reformer instincts coupled with environmental concerns and an interest in clean energy was bound to put Bass on a collision course with Don Young. The pair had already battled over the issue of special-interest financial provisions inserted into legislation at the last minute. The issue of "earmarking"—the often nontransparent, last-minute insertion into legislation of specific funds for pet projects—had come to a head very early in 2006 when Young tried to set aside $223 million for a bridge that would connect mainland Alaska to an island of about fifty residents. This controversial earmark became famously dubbed "the bridge to nowhere." When the national press exploded over the expenditure, the Republican Party was embarrassed.

Washington was abuzz. Something had to give. Bass and others began calling for an end to earmarking.

At a House leadership meeting in February, Young began yelling about Republicans who were thwarting him. People who saw the confrontation described Young as red-faced, furious, nearly out of control.

"This guy here single-handedly killed ANWR," Young shouted, referring to Bass's opposition to drilling in the Arctic National Wildlife Refuge. And now, here was this upstart junior congressman from New England again thwarting the "Congressman for All Alaska."

Young stomped across the meeting room. Standing in front of Bass, Young thrust out his fist. In it was a letter promising that the undersigned agreed that earmarks for projects in his district would be stripped out of an already passed transportation bill. "Why don't *you* sign this!" Young yelled, apparently believing he had backed Bass into a corner.

Young seemed certain that Bass, in an election year, wanted the gifts delivered to his home state. "Hand me the letter," Bass said. "I'll sign it." And he did. He was the only congressman to do so.

Within that context, Charlie Bass's decision to enter the seaside civil war was entirely consistent. Aware of the hysteria surrounding the Cape Wind proposal, he knew that congressmen from Massachusetts had been cowed into submission. There was no way they were going to raise Ted Kennedy's ire. Recognizing the national significance of Cape Wind, which would affect the entire New England electrical grid, Bass decided to take action. Two days after Bass issued his press release, Young filed his bill in the House of Representatives.

During the Easter recess, public outrage grew. In the context of the Washington lobbying scandals and the frustration over Iraq, Americans were looking for change. And they were looking for leaders with the courage to implement those changes. Cape Wind—and the brazen legislation intended to stop it—was an easy-to-grasp symbol of much that was wrong with American politics. Perhaps most important, the controversy was fairly easy to resolve: eliminate the egregious amendment.

The media were baying like a pack of Jack Russells after a fox. On April 14, the *Washington Times* called the anti–Cape Wind cabal onto the carpet using the most direct language yet. In an editorial headlined "Not-so-consumer-friendly Romney," the paper insisted the Massachusetts governor and presidential candidate would have to "abandon his unaffordable, outdated, self-serving NIMBY instincts. Otherwise, this former venture capitalist will send the unmistakable signal that he is more beholden to his wealthy political benefactors and the past than he is to middle-class energy consumers and the future." The editorial went on to name the Egan clan as major Romney financial backers, and ended by saying that "the man who opposes Cape Wind [Romney] had the nerve to declare that 'clean-energy technologies will reduce our dependence on foreign oil, air pollution and long-term energy costs.' He should have added: 'Provided that those clean-energy technologies are Not In My Back Yard, or in the front yard of my fund-raisers.'"

Bass then wrote a "dear colleague" letter, which he began circulating to House members. The letter contained strong language accusing Young of placing yet another earmark—an "anti-renewable energy legislative earmark"—into a bill. "Closed-door earmarks like these, inserted without a full debate and lacking an up or down vote, inappropriately circumvent the legislative process," Bass wrote.

"Dear colleague" letters are somewhat common in Washington, but they are only as good as their sponsors. Bass had a following of moderate Republicans but he needed the support of someone outside that circle.

He approached the then-head of the House energy committee, Joe Barton, and explained that Barton's committee, the House energy committee, had written 90 percent of the energy bill that finally passed.

How is it, Bass asked Barton, that the transportation committee can overturn our work? Last year, our energy committee—*your* energy committee—set rules about how offshore wind should be developed, and now an Alaskan heading the transportation committee is overruling that action. *Transportation*? How did that happen? And when it was put to him that way, Barton began to wonder . . . .

Barton is said to be a guy who cares about what the members of his House energy committee think about issues that relate to their district. He had three members from New England—Republican Charles Bass from New Hampshire, Democrat Tom Allen from Maine, and Democrat Ed Markey from Massachusetts. He polled his other two New Englanders. Markey stiff-armed the approach, making clear that he didn't want to touch the Cape Wind issue with a ten-foot pole. Given Kennedy's passionate opposition, Markey wasn't about to bite a hand that fed him.

Tom Allen of Maine, however, initially agreed to sign the letter. Allen comes from a largely coastal district close to the state's southern border with New Hampshire. Its voters are strongly proenvironment, strongly interested in clean-air issues, and very interested in welcoming business that would increase the use of Maine's ports—one effect of a healthy wind industry in New England, since wind-turbine components are usually transported on waterways.

When Allen agreed to sign (an agreement he later withdrew after having been contacted by project opponents) and when Markey's reluctance could be written off to timidity, Barton began to think. Maybe a line had been crossed here. Maybe he ought to be protecting all the hard work his committee had undertaken to get the energy bill through. After all, who knew more about energy: Joe Barton or Don Young?

Turf is an essential aspect of Washington politics. If Young could invade the energy committee's jurisdiction that way, who knew how far the invasion might ultimately go? Once Chairman Barton was engaged, the seaside civil war became quite a different kind of battle.

Joe Barton, a leader of the conservative wing of House Republicans, agreed to allow Charlie Bass to include the name "Barton" at the top of the "dear colleague" letter, right alongside Bass's own. Many who saw the two names coupled in this manner understood the significance of the temporary liaison. Opposition to the Young-Stevens-Kennedy amendment was no longer limited to the small group of Republican rebels led by Charlie Bass. The letter now had the stamp of mainstream conservative Republican leadership. Ultimately, only thirty-eight congressmen signed Bass's letter, a fact that Cape Wind did not make public, but the number wasn't as important as the names of some of the supporters.

After allowing his name to be used, Barton approached the Republican House leadership to complain that Young and Stevens had usurped the energy committee's power. Even worse, the Alaskans were doing a special favor for Ted Kennedy, of all people. This is not a problem the Republican Party needs to have, Barton warned.

Party leadership, apparently, agreed. Each day, project supporters expected to see the bill scheduled for a vote. Each day, nothing happened.

The Coast Guard bill lay underneath a pile of other, perhaps more pressing legislation. The more House leadership thought about that bill, particularly the way Bass had characterized it as yet another earmark, the less eager House officials were to move it forward.

The simplest way to handle such a problem is to decide not to handle it. Let sleeping dogs lie. They sent word through back channels to the Senate: This is your problem. It's a senator—Kennedy—and not a representative who wants the project killed. You created the problem. You solve it.

This was not good news for Alaskan senator Ted Stevens. At first, it hadn't seemed as though Stevens would have to pay much of a political price for his behind-the-scenes favor to Kennedy, who had, after all, taken him sailing on Nantucket Sound. Stevens expected to get the bill out of conference committee, take it on the chin from the press for a few days, then have the bill pass through the House and Senate and be done with the problem.

That wasn't what was happening. Instead, support for Cape Wind was exploding, while the effort to kill the project was imploding.

John Adams had conceived of the Senate, with its six-year terms that freed senators from the pressure of constant elections, as insulated from day-to-day chaos, but in the modern world, that is no longer the case. On the rare occasions when the public gets truly interested in an issue, the U.S. Senate becomes an important focal point. Its rules are more flexible, the number of officials to be swayed is much smaller, and its activities are more easily followed in the press and by the public.

While Charlie Bass held the House of Representatives at bay, Cape Wind's legions were flooding Senate offices with visits, phone calls, e-mails, and letters.

The day following the Cape Wind supporters' revival meeting on Nantucket Island, Gordon flew to Florida to visit his mother and his pregnant wife. No sooner had he arrived than he read in the paper that Republican senator John McCain was at a bookstore in a nearby city, signing copies of his new book.

Gordon immediately drove over, bought a copy of McCain's book and stood in line for ninety minutes. When his turn came, he knew he had only sixty seconds to get his message across. "Senator McCain, I'm Jim Gordon, CEO of Cape Wind, America's first offshore wind farm," he said, all in one breath as he thrust the books onto the table to be signed. "Oh yeah," said McCain. "I know all about that . . . ."

"Don Young has tacked an amendment onto the Coast Guard bill that will kill this project." Gordon said, pleading. "Can you help us?"

"I'm going to do everything in my power to fight this," Gordon remembered McCain saying. "Everything in my power . . . ."

Returning to Washington, McCain spoke to reporters in support of Cape Wind. News of Gordon's impulsive deed made the rounds of Cape Wind supporters and opponents alike. Supporters loved it, because it symbolized the kind of leadership style that made Gordon so different from someone like Doug Yearley or Bill Koch. Opponents groaned, realizing that it was awfully hard to counter that kind of unabashed zeal. Paid employees were unlikely to give up a vacation day just to have sixty seconds to beg a senator for help.

It began to dawn on project opponents that Gordon wasn't ready to go away, despite his dire circumstances. The Young-Stevens-Kennedy amendment may have emerged from the conference committee, but the fight was far from over. Appomattox was not just around the corner. Gordon was still at the head of his troops, leading them into battle with his standard flying.

Among those still fighting was Greenpeace's John Passacantando, who instructed his staff to develop a television ad focusing on the role played by Ted Kennedy in recent events. He wanted the ad to focus on wind energy. And he wanted it to make people laugh. The ad was intended to air in states with senators who might be vulnerable to Kennedy's personal pleas.

The staff came up with an animation showing a roly-poly senator standing knee-deep in saltwater, just off the Cape Cod shoreline. In the senator's hand was a wooden mallet. Sleek, modern wind turbines kept popping up out of Nantucket Sound, some to the right of the senator, some to the left. Each time, the senator whacked them back under the water. "I might see them from my mansion on the Cape," the Kennedy cartoon figure complained.

Greenpeace could only afford to air the ad a few times, but, as Passacantando had hoped, the ad hit the big time. The Greenpeace leader was called to a television studio to testify to a famous committee of two: Fox News's Hannity and Colmes. It was an experience Passacantando later likened to "running the bulls with your feet in cement shoes." During the interview, Fox News showed the Greenpeace ad. Commentators around the country started joking about the "eco-smackdown" between Greenpeace and Kennedy.

Greenpeace's animated depiction of Kennedy alerted the nation to backroom politicking in Washington, D.C.

Greenpeace's Washington staffers also invaded a Kennedy book signing. *America Back on Track*, the senator's first "policy" book in many years, was released in April almost simultaneously with the release of the Young amendment from its conference committee purgatory. Ironically, the book complained that "the disruption of oil supplies after Hurricane Katrina was a stark reminder of how vulnerable the economy is to sudden surges in energy prices" and chastised the Bush administration for having "no serious policy in place to . . . invest in new environmental technology." Among those technologies, the Kennedy book said, should be wind energy.

Greenpeace activists thought the senator was laying himself wide open with such statements, and that, given the widespread attention Cape Wind was getting, his book might make a splash, but not the kind he'd intended. "Do as I say, not as I do" is hardly an inspiring call to arms.

The book's publisher, Penguin, had planned a big release. The fun-loving folks at Greenpeace decided to help. They retitled the book, calling it: *How I Killed America's First Offshore Wind Farm*. The new title sat directly underneath a smiling picture of the old senator, dressed to the nines in polka-dotted suspenders and light blue pinstriped shirt. Staffers handed out the "book cover" as a flier to people standing in the long line waiting for Kennedy to sign their books. On the back of the flier was information regarding Kennedy's complicity in the backdoor amendment. When they got to the book-signing table, some people handed the senator their flier to sign instead of his book.

Not everyone saw the humor in Greenpeace's gag. RFK Jr. called Passacantando to tell him to leave the senator alone. "Your uncle doesn't get a free pass just because he's Ted Kennedy," Passacantando said.

Up in Boston, things were even hotter for the senator. Clean Power Now demonstrators formed a "human windmill" under the senator's Boston office windows. In response, Kennedy staffers called Gordon's Boston lobbyists, warning them away from the Kennedy office. The lobbyists explained that they had no control over Clean Power Now.

Demonstrators also attended the signing of the Massachusetts health-care-reform bill, which expanded health-care coverage to include thousands of people previously without adequate insurance. Passed by the state legislature and signed by Romney, Kennedy had no official role in the legislation, but everyone knew that, behind the scenes, the bill was his baby.

Standing beside Romney at the bill signing, Kennedy took credit for its passage. "After so many years of false starts, our actions have finally matched our words, and we have lived up to our ideals," he said.

Cape Wind supporters would have begged to differ with the senator on the matter of living up to ideals. They had not, however, been invited to attend the ticket-only event. Outside, though, Clean Power Now supporters picketed and chanted. When the elderly senator emerged from the signing, he walked quickly with his head down to his waiting SUV. Romney did not. Rather than face the picketers, he left by the back door, only to be yelled at by several Clean Power Now members who had staked out the governor's escape route.

Journalists peppered Kennedy with questions about Cape Wind. Right-wing commentators were having a field day, but the problem certainly wasn't limited to that group. To publicize his book, Kennedy gave an interview on National Public Radio—only to be asked about Cape Wind.

When John Keller, a political commentator for a Boston television station and a very outspoken opponent of what was now known commonly as the Kennedy-Young-Stevens amendment, approached the senator for an interview, Kennedy turned his back and walked away—very fast.

Keller, stunned, wrote: "You have to wonder if Ted looks so uncomfortable defending the rape and murder of the wind farm after years of clearing regulatory hurdles because he IS uncomfortable about his position, knows it's hypocritical, and wishes it—and pushy reporters hassling him about it—would simply go away."

The reporters would not, however, go away. Resentment was at a fever pitch. It was difficult for those who hadn't closely followed the Cape Wind battle to sort out the mess of conflicting claims, but it was easy to see that the project—and democracy—were being sold down the river using unsavory Washington-insider political strategies. Americans of all political groups were angry enough at Washington's politicians before Cape Wind made national news headlines. Gas prices were high. Home-heating prices were high. Rolling blackouts were occurring. The Middle East was a mess. And Washington politicians continued to live a life of luxury, taking free trips, eating free meals, drinking someone else's expensive wine, and generally enjoying themselves at the public trough.

Americans wanted change—big-time change, not just talk—and they wanted leadership. Jim Gordon and his ambitious Cape Wind proposal fit the bill. It had become a symbol of moving forward, out of the national energy morass.

The Coast Guard bill remained in limbo. Neither congressional body wanted to touch the issue. Then, in May, Senate energy committee heads Domenici and Bingaman ratcheted up the pressure. The duo wrote a May 3 letter to the Senate Majority and Minority leaders, Bill Frist and Harry Reid, that said, "It is particularly objectionable to subject the development and implementation of national energy policy on submerged federal lands to the caprice of an individual state governor."

Particularly significant to Washington insiders was the statement that the pair would "oppose allowing the conference report to proceed to a final vote on the floor of the United States Senate, unless both the Senate and the House of Representatives pass a correcting resolution" that removed the onerous gubernatorial veto.

Domenici, who had been ready to give up when the bill emerged from conference committee, was back on board. Both senators were willing to go to the mat over Stevens's amendment. At a wind-energy conference in Washington, Leon Lowery, Bingaman's top energy adviser, told conferees that the senator would filibuster the Coast Guard bill if necessary.

One explanation of Domenici's return to the battle was the claim made by both Kennedy and the Alliance to Protect Nantucket Sound that Cape Wind had received special, lobbyist-engendered treatment at the hands of the energy committee. The Energy Policy Act had clarified that the Cape Wind project would need to be reviewed by Minerals Management Service—but would not need to go all the way back to square one. Kennedy and others had tried to require competitive bidding for the site, but Domenici's and Bingaman's staffs believed that since Cape Wind had been in the pipeline for five years, the project should be grandfathered to a degree.

Outraged project opponents accused Domenici and Bingaman of behind-the-scenes dirty dealing. For Domenici in particular, the charges, which Kennedy repeated frequently for months, amounted to a call to arms.

Cape Wind supporters needed to encourage Senate Majority Leader Bill Frist not to be overeager in calling the bill up for a vote. Gordon hired a public-relations firm in New Hampshire to visit that state's newspapers to explain the situation. New Hampshire's key role in presidential primaries (brought about in part by Charlie Bass's grandfather) meant that Frist, who hoped to run for president in competition to Romney, would care a lot about what New Hampshire voters thought about his actions. When New Hampshire journalists also began to write about the Cape Wind ambush, elected officials again saw that the project had widespread support.

On May 4, Bush administration under Secretary of Energy David Garman weighed in with a letter saying that the "Administration would like . . . to express its concerns" about the amendment. The letter also said that "the New England Independent System Operator is counting on this project and the power it will provide." It ended by expressing the concern that the chaos caused by Cape Wind opponents was having a seriously detrimental impact on the emergence of offshore wind throughout the United States.

Funders of the Alliance to Protect Nantucket Sound struggled valiantly against this onslaught, but they began to look more and more like the Gang Who Couldn't Shoot Straight. Everything they tried either ended up as a joke or raised the hackles of yet another interest group. Alliance executive Charles Vinick and Ernie Corrigan were seen walking the halls of the Capitol, visiting a number of officials. Vinick admitted having spoken at length with Young months earlier.

On May 4, the same day that Garman's letter was issued, the Alliance placed an ad in *Roll Call,* the daily Capitol Hill newspaper. "Why is this man smiling?" asked the ad. It featured a hand-drawn caricature of an oversized Gordon splashing through Nantucket Sound, stomping on yachts. The character clutched oodles of cash in both fists. Bills dripped from the greedy man's pockets. On the sole of one shoe was a dollar sign. Wind turbines lurked in the background. Unfortunately for the Alliance, some found the cartoon offensive rather than funny. Some claimed the caricature was anti-Semitic.

Despite Alliance lobbying attempts, the pressure on the anti–Cape Wind senators intensified. A Robert Novak column on May 4 ridiculed Kennedy and called Young and Stevens "reigning princes of pork on Capitol Hill." Overall, it was a somewhat remarkable situation. Robert Novak and Greenpeace were on the same side. The energy developer had become the guy with the white hat, and Kennedy, the selfish autocrat.

It was more than the Massachusetts senator could bear. On May 9, both Stevens and Kennedy won a few minutes of time to speak on the floor of the U.S. Senate. First came Stevens. Nantucket Sound, said the senator, is special. People sail thirty-foot sailboats there, he explained, and the towers will be much bigger than the sailboats. Stevens showed himself woefully uninformed. Nantucket Sound, he repeated several times, is "the proposed site for the largest and most expensive offshore wind energy project ever undertaken in the world." The truth was that Cape Wind was much smaller than projects under development elsewhere around the globe. The London Array, for example, first proposed in 2001, will consist of 271 wind turbines with a capacity to produce as much as 1,000 megawatts. Nonetheless, not one to be held to facts, the senator from Alaska repeated this assertion several times. "Congress . . . ought to listen to the senior senator" from Massachusetts, Stevens said.

Kennedy was next up. The senator's highly emotional ten-minute performance was rather sad. It was all too apparent that, after five years of vehemently opposing Jim Gordon's idea, he still did not understand the project's basics. Much of what Kennedy said verged on the ridiculous. "It is completely exempt from all of the protections! Completely exempt . . . . Oh no, not for them, they have a special provision. . . . Hello?! . . . the EPA has said that this project is completely inadequate. . . . Do you think they have to make any corrections? Absolutely not! . . . . To get the sunlight!—the sunlight!—of consideration by the state as well as by the other agencies so that they can make a comment on this, which they are not permitted to do under the special provisions of this law . . .

WHY IS THIS MAN SMILING?

When Congress prepared to pass the Energy Policy Act of 2005 last summer, Cape Wind's friends in Congress slipped in a special-interest amendment that gives this wealthy energy developer a deal that most developers can only dream about: a no-bid, non-competitive deal for development rights on 24 square miles of Nantucket Sound for the largest industrial, offshore wind energy complex in the world.

Imagine being handed over 15,000 acres of ocean just off the coast of Cape Cod and the islands of Martha's Vineyard and Nantucket. No standing in line. No messy competition to spoil your chances. One developer, and the deal of a lifetime.

YOU'D SMILE TOO IF you had grabbed the sweetheart deal that Jim Gordon and Cape Wind got from Congress last year.

But wait, this deal gets better.

Cape Wind also qualifies for over $300 million in federal Production Tax Credits. That's $30 million a year for 10 years. And they would also qualify for an estimated $50 million a year in Massachusetts green credits for the lifetime of the project. That's another $1 billion over a projected 20-year project life. They even get extra tax benefits by being able to depreciate this billion dollar project over just 5 years instead of the usual 20.

What do Cape and Islands taxpayers get after handing over 24 square miles of public trust ocean lands to this developer? We get an industrial-scale project that will displace our commercial fishery, endanger boats and airplanes, harm the environment and risk a regional economy that relies on the beauty of Nantucket Sound.

Join with the many elected officials, fishermen, ferry operators, airports, chambers of commerce, taxpayers, and organizations who oppose destroying a national treasure for the corporate exploitation of one developer.

Protect Public Safety, not Private Exploitation of Public Lands.
Support the Coast Guard and Maritime Transportation Act of 2006.
Visit www.saveoursound.org or call 508-775-9767 to help.

and the state says this is a fragile area. . . . It's not right! It's not right, Mr. President! We deserve to have at least fairness in this. . . . This has been declared, for four hundred years it's been used. . . ."

My, but the man was angry. Kennedy attacked Gordon on the floor of the Senate, all but using his name. He stopped just short of accusing Gordon of corrupt practices. Later, he would compare Gordon to Halliburton, which had recently been in the news for misuse of funds in Iraq.

Gordon didn't know exactly what to make of Kennedy's performance. Over the past several months, he had come to think of the Senate as a kind of New World Kabuki Theatre, where ancient and ritualistic plays were performed that made sense only to the fully initiated.

The old senator's all-too-public paroxysm of mania was to be the swan song of the Stevens amendment. Up in Boston, Romney was continuing to insist that he would stand by his man on the Senate floor, but for most, even for Senator Edward M. Kennedy of Hyannisport, Massachusetts, the pressure had just become too great to bear.

Behind the scenes, Domenici and Bingaman, more and more indignant over Kennedy's charge that Jim Gordon and Cape Wind had received special, "completely exempt," treatment from the Senate energy committee, refused to back down. They made their position clear. Lengthy backdoor negotiations with Kennedy were under way. There was no way, they explained to Kennedy, that the odious amendment would make its way through the labyrinthine halls of senatorial process. The press, the public, grassroots groups like Greenpeace—all had converged to bring Cape Wind to the forefront, and it wasn't going to be possible for Kennedy to receive this favor.

Stevens, Young, and Kennedy had gone for broke to stop Cape Wind, but had lost. How offshore wind would be incorporated into the American energy framework was something that Bingaman, in particular, cared about a great deal. This his staff explained as gently as possible to Kennedy's staff.

Conversations among the four senators and their staffs dragged on for several more weeks, while the senior senator from Massachusetts prepared himself to accept defeat gracefully.

On May 26, the *Boston Globe* ran a front-page story—right next to a picture of the just-convicted Ken Lay, former CEO of the infamous Enron energy company—headlined "Kennedy eases his stance on wind plan." The story explained that, according to a Kennedy aide, "the Massachusetts Democrat would still prefer that the governor . . . have the final say, but realized that's not politically feasible in this legislation."

The stalemate seemed, after nine months, to be coming to a close. But still nothing happened. Behind the scenes, Kennedy was struggling with all his might to get what he wanted. On the other hand, Domenici, fueled by his anger over having been accused of behaving corruptly in terms of Cape Wind, was not going to give in.

Finally, on June 20, a "correcting resolution" was filed in the House of Representatives, under Don Young's name. The new language was a sound, clear defeat for Cape Wind opponents. It said only that, in regards to "an offshore wind energy facility in Nantucket Sound," the Coast Guard Commandant would specify the "reasonable" terms and conditions required to ensure navigational safety. The Senate's energy committee issued an official press release the following day explaining the compromise. Domenici and Bingaman also issued statements. Bingaman's statement allowed Kennedy to save face and called the new amendment language "appropriate."

Domenici was not so kind. The energy committee chairman said the outcome preserved "the integrity of the siting procedure outlined in the Energy Policy Act . . . . We will use the siting model we created in the energy bill. That's a sound model . . . . It prevents local special interests from torpedoing a reasonable and much-needed energy project in federal waters."

Word of the resolution reached Jim Gordon just before he rose to speak at a renewable-energy financing forum at New York City's Waldorf Astoria. "We now get to hear from one of the media stars of renewable energy, Jim Gordon," said the moderator. "He is easily one of the most tenacious guys I've ever come across . . . ."

At the podium, Gordon looked like he was back boxing at summer camp with the boy who ultimately died committing armed robbery. He danced on his toes. Weaving back and forth, filled with adrenalin, he looked like Rocky Balboa. He had come a long way from five years earlier, when he sat by himself, ramrod straight, as the Alliance beat their drums in the Barnstable Town Council meetings.

Speaking without notes, Gordon began to relate the whole nine-month political ordeal, from Young's placeholder amendment to the most recent days. The more he talked, the more his hands waved in the air. His cadences sped up. The adrenalin kept flowing and his hands flew above his head. He was in a manic mode, like a soldier relating stories from the field of battle. By the time he got to the punch line, he was breathless. "Two hours ago . . ." And he told his audience about the resolution.

As a wave of applause swept the room full of hundreds of people, Gordon kept talking. He wasn't taking bows by himself. "We would not have received this press release," he cautioned the audience, "if it wasn't for many people in this room today, as well as bringing together of labor, the environmental community, most important, the citizens of Massachusetts."

Later, Gordon's team celebrated, too.

"We can speak with a powerful voice when we get it right," said John Passacantando later. "When we get it right—that's a beautiful thing."

Epilogue: Jim Gordon's Big Idea, Redux

In late July 2006, Democratic congressman Bill Delahunt stood beneath a canvas tent in Hull. He was orating again, this time at the "dedication ceremony" of the town's second wind turbine. The machine had actually been generating for months. The fancy ceremony was just a face-time opportunity for Delahunt and Forest to take public credit for the work done by private citizens like energy activist Malcolm Brown.

After his speech, Delahunt issued a press release that attempted to connect Delahunt with a $1.7 million forgivable loan allowing Hull to study the construction of another four turbines. In fact, the money came from the Massachusetts Technology Collaborative, the state organization established through the efforts of Matthew Patrick and many other renewable-energy advocates.

Delahunt's oration notwithstanding, in Massachusetts, the future of wind energy looked grim. Over the five years of the seaside civil war, under the Romney administration, only three commercial-scale wind turbines had been built, totaling only about 3 megawatts. By comparison, the Cape Wind project had a nameplate capacity of 468 megawatts. Elsewhere in the United States, by the time of this writing, well over

10,000 megawatts of wind energy existed, and the American Wind Energy Association, the industry lobbying group, predicted that more than 3,000 new megawatts would be built in 2007 alone.

How much of that innovation would be allowed into New England remained unclear. Anti-wind-energy forces remained on active duty, doing whatever they could to stall development. Over the summer, for example, Alliance head Bill Koch peddled one of the most ridiculous anti–Cape Wind newspaper columns to have been written to date. To electrical grid managers, who understood the intricacies of the power markets, many of Koch's claims were about as cogent as his comparison of wind turbines to striptease artists. In his piece Koch claimed that if the project were built, "Cape Cod residents would have to pay $1,300 more per year for their electricity." The source of this absurd figure was never revealed. When Alliance witnesses made similar claims before the state's Energy Facilities Siting Board, that board found instead that the project would actually decrease New England rate-payers' costs by roughly $25 million a year. Koch also claimed that Cape Wind would have to build a second fossil-fuel power plant to accompany the wind plant, since wind does not blow continuously. New England grid managers had said repeatedly over the past five years that no such plant would be needed. Instead, they said, the power would be easily absorbed into the system, which, they said, is constantly required to compensate for the ebb and flow of electricity.

Unfortunately, silly claims like those of Bill Koch had an effect on some readers who did not yet understand that in today's competitive electricity markets, cost and price are no longer the same thing. If the project's product—a kilowatt-hour—were too expensive, there would be no buyers. That's because, in today's deregulated markets, there *are* no guaranteed buyers.

Koch's column, complete with its unfounded claims, appeared in, of all places, the *Wall Street Journal*, conferring upon his flimflammery a certain legitimacy. Few newspaper readers understand that opinion columns are rarely held to high standards of accuracy by newspaper editors. Indeed, at the end of the column the paper ran a tagline identifying Koch as a company president, holder of a doctorate from MIT—and, of course, 1992 winner of the America's Cup. But editors did not tell their readers that Koch was the head and a major funder of the group trying to kill Cape Wind.

Mark Forest continued his efforts, too. Since the June defeat of anti–Cape Wind legislation in Washington, he had been hard at work creating the "Massachusetts Offshore Renewable Energy Development Act," a document of just over two pages purporting to be a legislative white paper. If passed by the state, Forest's ideas would severely limit innovation in Massachusetts, a state that once touted its role as the nation's scientific and technological leader. Under Forest's pen, the legislation would allow only certain specified saltwater locales to host wind turbines or any other experimental clean-energy technology. Before those areas were chosen, local communities would have to agree to accept the project. Since Forest's "legislation" appeared to have been designed to protect property values, the end result would of course be that poor communities like Fall River would be chosen for the experiments while wealthy communities like Osterville could choose to be left alone.

"Isn't that beautiful?" muttered Jim Gordon, on seeing Forest's work praised by *Cape Cod Times* editors.

For his part, Ted Kennedy continued telling people that Nantucket Sound was a "marine sanctuary," that his family's favored sailing waters were "pristine," and that Nantucket Sound, replete with its private beaches open only to the rich, ranked right up there with national parks like the Grand Canyon.

Meanwhile, managers at New England's ISO were writing yet another report warning that the region needed to build 5,000 megawatts of power generation over the next decade. Given that the region already had an overdependence on natural gas, the report warned, some other source of fuel would have to be used. As an option to wind energy, the report suggested the plants might be nuclear or coal powered. Although *Cape Cod Times* editors had also suggested a nuclear-power plant, such a plant was unlikely to be built in Osterville, Massachusetts.

Conversely, a liquefied natural gas facility was a possibility. One had already been proposed for Fall River.

Five years earlier, in late 2001, Jim Gordon and his staff at Energy Management Inc. proposed a bold idea that, they believed, would change the face of New England's energy infrastructure. Once the public realized the scale of the energy revolution at hand, Gordon and company believed, people would clamor for the new technologies and buy into his proposal, just as they had bought into his salesmanship in college when he sold cable television door-to-door. Talk long enough, and patiently enough, Gordon believed, and people would, eventually, come around to your point of view.

He hadn't reckoned on the Nantucket Sound crowd, who had settled into the coastline nearly one hundred years earlier and weren't about to give up ground, no matter how useful the retreat might be to the future of America.

Whether or not Gordon's project will ever be permitted remains at the time of this writing quite uncertain—although if you say that to Gordon, he immediately counters: "Not *if* the permits are granted. *When* the permits are granted."

If Cape Wind *is* built, observers expect Gordon and his investors to make quite a bit of money. Washington legislators, including Ted Kennedy, have approved a variety of tax incentives to encourage clean-emerging-energy technology. The cost of project construction—estimated right now to be roughly $1 billion—can be depreciated over five years, instead of the more typical ten. A production tax credit of nearly two cents a kilowatt-hour is available, and Massachusetts has created a "green energy" market expected to average roughly two and a half cents for each kilowatt-hour produced. Does that make Gordon and other investors greedy?

"These are incentives that were set up by Congress, not by us," said Gordon. "Is somebody 'greedy' because they qualify for a program?" If so, then why did the federal government pass such legislation? Why does Congress, including Ted Kennedy, extend the production tax credit? In the fossil-fuel world—that inhabited by Bill Koch, Doug Yearley, and many other Cape Wind opponents—tax incentives are as common as sand at the seashore.

What the future holds for offshore wind in America, once upon a time a promising technology, also remains uncertain. Since Gordon proposed his idea five years ago, as of this writing in February 2007, not one wind turbine has been built in American waters. The massive onslaught of lies perpetrated by Cape Wind opponents has frightened off less tenacious developers or publicly traded companies, which cannot afford the bad publicity.

Money and corrupt government officials are hijacking our nation's economic and environmental future. At the beginning of the seaside civil war, an Irish wind-power developer called Airtricity came to the Boston area to establish its U.S. offices. It seemed to the Airtricity executives that the Boston region would be a great operational base. The city had a reputation for forward-looking, liberal-thinking ways. The region seemed to respect the environment and to care about clean air and clean water.

But as the Irish executives watched the Cape Wind drama unfold, they began to have their doubts. "It turned out that it was quite the opposite from what we'd expected," said Martin McAdam, head of operations and new business ventures for Airtricity. "At every opportunity, people got in the way of projects. Watching the vested interests opposing Cape Wind left a distasteful taste in our mouths. The Massachusetts governor was obstructing projects, while Texas was welcoming them. It's quite remarkable sometimes, the difference between the things people say and the actions they actually take."

McAdam said that the Romney administration never welcomed the Irish company into Massachusetts, never reached out to them or even invited them for lunch. "Let me be honest with you," McAdam said, "on balance, there was tremendous support from the Massachusetts community for Cape Wind. We thought that meant that democracy should work. If the people are supporting this, the elected leaders should be supporting this. I wouldn't like to speculate as to what's behind Governor Romney's motivation, but from our perspective, Nantucket Sound is one of the best offshore locations in the United States. But the business environment in Massachusetts just does not support wind development. Governor Romney has made that clear."

In 2003, Airtricity, which expects to do at least $2 billion in business over the next five years, moved its U.S. offices to Chicago.

Minerals Management Service began its own round of permitting hearings for Cape Wind just as the Alliance-caused chaos in Washington was winding down. At one of the public meetings, Ernie Corrigan showed up with a bus full of Alliance supporters, who, as usual, signed up very early in order to be able to dominate the first hours of the public hearing.

Before the meeting, the Army Corps' communications officer Larry Rosenberg, who was there as an observer, stepped outside for a cigar. There, he found Ernie Corrigan. "Well Ernie," Rosenberg said, "here we are again. It's five years now. You've been at this a while."

"Yep," Corrigan smiled, "I've been riding this horse a long, long time." Corrigan never revealed how much the Alliance paid him over the years to play his own part in the public discussions over the future of wind energy in the United States. Best estimates were that, on a yearly basis, he earned well into the six-figure range.

Authors' Notes

Author's Note from Wendy Williams

This is an old story, dressed up in new technology. It may, in fact, be as old as the primate brain, obsessed as it is with power and social status. I watched it unfold from the very beginning, writing about it frequently for *Wind Power Monthly,* the international wind-industry newsmagazine published in Denmark.

Because I live on Cape Cod, I was able to attend the initial meetings in person. I had the opportunity to listen to presentations and to speak to people on both sides of the issue. In addition, I had the opportunity to attend the Wianno Club strategy meeting that followed Gordon's announcement. I was struck by the intent of those at the meeting to stop this project in its tracks, even before the facts of the case had been clearly laid out.

For quite a while, I expected to see the various parties sit down together to negotiate. But the more I talked with opponents, the more I realized that battle lines had already been hardened. From the outset, there was no room for discussion. When I learned that Ted Kennedy had refused to meet Gordon for even a few minutes, I understood that compromise was not in the cards. Larry Wheatley made this doubly clear when I spoke to him at one of the many court hearings that occurred following the election won by Matt Patrick. "Not one turbine in Nantucket Sound!" was Mr. Wheatley's war cry.

I did not at first see Cape Wind as a subject for a book. It seemed like a very local story without great national interest. I changed my mind after attending the public hearing on Martha's Vineyard that opens this book. At that point, I understood quite clearly that Marathon Oil's Doug Yearley firmly intended to carry through on his threat to use all the money necessary to manufacture what was supposed to be a "grassroots" movement.

In my thirty years as a journalist, I had never seen such a brazen attempt to obstruct the democratic process. Unfortunately for project opponents, their behavior was often so blatantly over-the-top that, with each bungled effort, more people took notice.

A genuine grassroots movement did indeed grow up, but it was not the grassroots movement paid for by Doug Yearley and other wealthy project opponents. The grassroots movement consisted of people offended by the hijacking of the democratic process. Many of them, like me, were not necessarily supporters of Cape Wind per se. They were people who believed that American democracy was an important value, one which it was their democratic duty to protect.

To me, this is a good-news story, a tale of all kinds of people who behaved patriotically, who stood up to defend their government. These people believe that if some valid reason surfaces showing that Jim Gordon's project is not a good idea, then the project should not be built. But the decision should occur in an open and aboveboard manner—and not behind closed doors in Washington or over the dinner tables of Oyster Harbors and Nantucket Island.

Toward the end of writing this book, I attended a dinner at the Harvard Club up in Boston. My dinner companion was a cultured and courtly older gentleman. "I have a house in Woods Hole," he told me, "and a house in Nantucket. I sail my boat from one to the other. Everywhere I sail on Nantucket Sound, I will see these windmills."

"Yes," I answered, "I think you will."

"Well, I don't want to," he said.

"And you have a perfect right to that opinion," I answered.

And he does. But his opinion shouldn't count for any more than an opinion from someone who lives on West Island and doesn't want fuel oil on her beaches, or an opinion from someone who lives in poverty-stricken Fall River and doesn't want emissions from a coal plant covering her dining table every day.

I grew up in the coal country of Western Pennsylvania. When I was a child, Pittsburgh was a bleak world. The poisons in the air from the coal plants and steel mills kept nature at bay. Greenery was limited. Slag heaps were plentiful, though.

My cousins still live in New Florence, Pennsylvania, a small and economically hard-hit town in the Appalachians that has recently become surrounded by no fewer than five new coal-fired power plants. To remove the coal ash from her windows, my cousin Diana has to use steel wool.

Most of the power in those plants is sent to the East Coast. My cousin, not surprisingly, thinks wind turbines are great. She also thinks people on the East Coast should make their own electricity and leave her windows alone.

It should be up to the American people—and not a small, tight, well-connected, and well-funded network in Washington—to decide how much weight to give to Diana's opinion and how much weight to give to that of the wealthy yachtsman.

The Chinese have a saying: "The mountain is high, and the Emperor is far away." By this, they mean that it's one thing to have a law—and quite another to faithfully uphold that law. In the United States, theoretically at least, we have no emperor. In his place, we have the American people.

It is we ourselves who are responsible for seeing that we remain a nation of laws and a nation that respects the democratic process. If we are not vigilant, no one will do it for us. If we become a nation ruled over by a small and wealthy elite who pay people like Ernie Corrigan and the lobbyists in Washington to manipulate the democratic process in order to get their way, we have only ourselves to blame.

Author's Note from Robert Whitcomb

With some of my father's family well ensconced on the Cape since the 1600s, I had a strong, if increasingly erroneous, sense of the place as I thought about it while living in various places as far away as Paris. My memories of the Cape and Islands were of a mild and rural land, where droll relatives (yes, they really were) lived in old houses with cedar siding silvered by the sea air, where the elms and oaks shadowed the narrow roads and there really was a general store. (The one in the village where my paternal grandparents lived was owned by a man called George West; he overcharged the summer people but extended credit to the real locals, and would visit the houses of the old people if they failed to show up at his store for their routine shopping.) A trip to fabulous downtown Falmouth to patronize a clothing, book, or hardware store was an outing. For the exotic, we'd go see the weird scientists and aquarium fish in Woods Hole, the famed oceanographic center, down the road.

Old Cape Codders, including my relatives—some of whom were descendents of the Quakers who were a big part of the original English settlement—were not particularly happy with change, social or otherwise. A Cape Cod aunt told me that when she introduced the man who would become her husband to one of her uncles, a Woods Hole character, during World War II, his response to the fiancé's last name was to ask, "What kind of a name is that?" and huffily walk off.

The Cape was once a refuge for me. My immediate, thermo-nuclear family lived in an affluent Boston suburb where some of the activities recalled a Eugene O'Neill play. The disorder and unpredictability of this household was such that any visits to the calm (if sometimes bossy) affability of my grandparents' Cape Cod house were joyful. It seemed a sweet world—as Henry James said of the Cape: "mild and vague and interchangeably familiar with the sea"—the sea being a potent sedative, to live by if not always to travel on.

Then it changed, as such sweet places along the shore tend to do, a victim of its beauty, of its proximity to megalopolis, and of the fact that by New England standards its climate is tropical; you can often play golf in the middle of winter. Before the change, my family had traveled to Cape relatives' gray houses on a narrow, undivided road with vegetable stands and Burma-Shave signs. Then the extension of big divided highways to the Cape Cod Canal bridges in the 1960s brought many more cars, and a massive building boom that included subdivisions, malls, and the other ambiguous charms of suburban life. The big roads made it faster for cars to reach the Cape and the ferries, the Islands—until the lure of "fast" highways drew so many cars that city-style gridlock became common on some roads, first in the warmer months and then year round. (The rich increasingly avoided the traffic problem by flying to the Cape and Islands, often in their own planes. John F. Kennedy Jr. met his death on his way to the Vineyard from New Jersey, inexpertly piloting a small plane through a pollution-thickened haze.)

Meanwhile, in the past several decades, the affluent have rapidly accelerated their occupation of, and sealing off of, the shore, aided by seventeenth- and eighteenth-century laws defining coastal ownership. A couple of my luckier relatives enthusiastically supported the sealing off.

There had been, of course, plenty of rich summer people in the region before the new highways started funneling throngs onto the Cape bridges in the 1960s, but the summer wealth on the peninsula exploded with the bigger roads, expanded air travel, and vast new fortunes in the high-tech, financial-services, and real-estate booms of the 1980s and 1990s. The Cape and Islands were for sale to the highest bidder, and there were plenty of bidders. And many of the rich newcomers wanted to live near other movers and shakers (which affirmed each other's importance) and to display their glory by building huge houses—

although these structures didn't always sit gracefully in the region's soft, low terrain.

When ruminating about the Cape, I started to see less glittering sea and more marinas and "No Trespassing" signs.

And when I did visit, I noticed how the old town centers, with their quirky shops and Quaker meetinghouses and white congregational churches, had lost their community magnetism, as they were bypassed by shopping malls on the edge of town. The Cape and Islands' small-town life became exurban and then, at least on the Cape, even suburban—that is, suburban in search of a city. Actually that city appeared in the unsightly form of Hyannis, whose Los Angeles–style sprawl was right on the sound called "pristine" by foes of the Nantucket Sound wind farm.

Of course, many of these changes were happening elsewhere on America's coasts, but life on the pre-exurban, pre-glitterati Cape Cod and Islands was so nice that the changes seemed particularly sad there, at least to me. The plowing up of, say, the Jersey Shore, so close to New York City and Philadelphia, somehow seemed less heartbreaking. But then, that's probably just my provincialism and selfishness.

The transformation of the Cape and Islands has not been a pretty thing, but socially, aesthetically, economically, and environmentally, it has been fascinating—a microcosm of American life. And the reaction to the plan to put 130 huge wind turbines in the middle of Nantucket Sound brought it all together.

Acknowledgments

Many people deserve to be thanked for their patience and for their willingness to explain the complicated technology and science that provided the foundation for our understanding of these issues. In particular, thanks to Chris Reddy of the Woods Hole Oceanographic Institution, who let us accompany him on his oil spill research forays and who spent hours explaining the chemistry of oil; to George Woodwell, who founded the Woods Hole Research Center; to Gordon van Welie, Stephen G. Whitley, and others at ISO New England who helped us understand how the electric grid works; and to Al Benson of the Department of Energy, who explained a great deal about New England's energy crisis and who is a fount of interesting and accurate information about energy in today's world.

Thanks also to Tom Wineman of Clean Energy Design and Clean Power Now, who shepherded us around Nantucket Sound on his sailboat, the *Blue Goose,* and to Charles Kleekamp of Clean Power Now, a retired engineer who struggled to research the technological facts surrounding Cape Wind. Democracy needs more heroes like Mr. Kleekamp and his wife, Kathryn, a retired scientist. Thanks to Larry Rosenberg of the U.S. Army Corps of Engineers, who helped provide basic information on the facts of the proposal and the permitting process as well as the behind-the-scenes tales of the various corps public meetings in this book. Thanks to Lyn Harrison of *Wind Power Monthly,* the industry magazine, who is a fount of information on the wind industry and willingly explained many of its details.

Thanks also to Russ Haydon of Osterville, a retired man who stood up and spoke out for the democratic process, even though he trembled as he did so. Mr. Haydon is in this book because he stands for the common man or woman, the average American. He believed in America so strongly that he spoke out, even though no one paid him to do so. These people make America work.

Additionally, others deserve to be thanked for their willingness to explain, on background, the behind-the-scenes political maneuvers that occurred in the course of the five years. They did this sometimes at great risk to their careers. They were particularly helpful on the incidents occurring in Washington.

We would also like to thank the many libraries that helped us research the historical documents that provide the foundation for the background we used to set this battle in its historical context. The librarians at the Cotuit Public Library, without ever once complaining, found books from locales as far away as London. We would also like to thank the Howard Gotlieb Archival Research Center of Boston University, where we found the papers of Edward C. Stone; the Department of Special Collections at the Charles E. Young Research Library of UCLA; the American Antiquarian Society located in Worcester, Massachusetts, which has a delightful book written by the wife of Moses G. Farmer and many local newspapers of that era; the Pittsburgh Historical Society, which provided much of the background on the Mellon family; the Osterville Historical Society, which provided documents on the history of that village; and the Cape Cod Community College's historical archives, which provided access to its plethora of documentation on the history of tourism in the region.

Thanks also to the many organizations that held conventions and educational seminars and allowed us to attend, as guests. Those organizations include the American Wind Energy Association, the National Hydrogen Association, and Law Seminars International.

This book is the result of hundreds of hours of sitting and listening to both sides of this conflict at the seemingly endless numbers of public meetings that have occurred on Cape Wind since October 2001. Although none of the project opponents or the employees of the Alliance to Protect Nantucket Sound agreed to be interviewed specifically for this book, we nonetheless feel very well acquainted with their point of view, having spoken with them directly and listened to them frequently throughout the course of the five-year conflict.

As will be obvious to the readers of this book, Jim Gordon, Dennis Duffy, Mark Rodgers, and Rachel Pachter endured a yearlong onslaught of questioning and requestioning. We thank them for their patience.

We must also thank our agent, David Miller of the Garamond Agency, who understood the importance of this project from the beginning. And thanks to our editor, Lisa Kaufman, who brought this project to fruition.

Particular thanks go to our patient spouses, who had their own special roles to play. Nancy Spears, an accomplished artist in her own right, did the lovely drawings. Greg Auger stalwartly attended countless hours of meetings and gatherings and provided many of the photos used herein.

One final note: Everyone asks whether Jim Gordon's Cape Wind proposal will ever be built. We are reluctant to predict the outcome. Jim Gordon is a very persistent energy entrepreneur. The number of people supporting the project in Massachusetts seems to be very high, if you can believe polling numbers, some of which say project support ranges around 80 percent throughout the state. On the other hand, while the pure numbers of project opponents on Cape Cod is dwindling, the determination of the core group of opponents is not. And, as this book makes clear, they have plenty of money and political support that ensures that their opposition is heard.

Bibliography

A Partial Bibliography

Adams, Edward Dean. *Niagara Power—History of the Niagara Falls Power Company, 1886–1918* Niagara Falls, NY: Niagara Falls Power Company, 1927.

Allen, Frederick Lewis. *Only Yesterday*. New York: Harper & Row, 1951.

Auchincloss, Louis. *Theodore Roosevelt*. New York: Times Books, 2002.

Baldwin, Neil. *Edison: Inventing the Century*. New York: Hyperion, 1995.

Barnstable, Town of. *The Seven Villages of Barnstable*. Barnstable, MA: Barnstable, 1976.

Beaudreau, Bernard C. *Mass Production, the Stock Market Crash, and the Great Depression: The Macroeconomics of Electrification*. Westport, CT: Greenwood Press, 1996.

Beebe, Lucius. *The Big Spenders*. New York: Doubleday & Company, 1966.

Belfield, Robert Blake. *The Niagara Frontier: The Evolution of Electric Power Systems in New York and Ontario, 1880–1935*. PhD dissertation, 1981. University of Pennsylvania.

Beran, Michael Knox. *The Last Patrician: Bobby Kennedy and the End of American Aristocracy*. New York: St. Martin's Press, 1998.

Bodanis, David. E = MC^2: *A Biography of the World's Most Interesting Equation*. New York: Berkeley, 2000.

———. *Electric Universe: The Shocking True Story of Electricity*. New York: Crown, 2005.

Bradford, Sarah. *America's Queen: The Life of Jacqueline Kennedy Onassis*. New York: Viking, 2000.

Brands, H. W. *The First American: The Life and Times of Benjamin Franklin*. New York: Doubleday, 2000.

Bright, Arthur A., Jr. *The Electric-Lamp Industry: Technological Change and Economic Development from 1800 to 1947*. New York: Macmillan Company, 1949.

Brower, Kenneth. *Freeing Keiko: The Journey of a Killer Whale from Free Willy to the Wild.* New York: Gotham Books, 2005.

Brown, Andrew H. "New St. Lawrence Seaway Opens the Great Lakes to the World." *National Geographic Magazine*, March 1959.

Brown, Sanborn C. *Benjamin Thompson, Count Rumford*. Cambridge, MA: MIT Press, 1979.

Brush, Edward Hale. "Electrical Illumination at the Pan-American Exposition." *Scientific American* Supplement, January 19, 1901.

Bryce, Robert. *Cronies: Oil, the Bushes and the Rise of Texas, America's Superstate*. New York: Public Affairs, 2004.

Burkett, Elinor. "A Mighty Wind." *New York Times Sunday Magazine*, June 15, 2003, 48–50.

Burrows, Edwin G., and Mike Wallace. *Gotham: A History of New York City to 1898*. New York: Oxford University Press, 1999.

Butler, Karen T. *Nantucket Lights: An Illustrated History of the Island Legendary Beacons*. Nantucket, MA: Mill Hill Press, 1996.

Butti, Ken, and John Perlin. *A Golden Thread: 2500 Years of Solar Architecture and Technology.* New York: Van Nostrand Reinhold Company, 1980.

Carpenter, Edmund J. *The Pilgrims and Their Monument*. New York: D. Appleton & Company, 1909.

Cathcart, Brian. *The Fly in the Cathedral: How a Group of Cambridge Scientists Won the International Race to Split the Atom*. New York: Farrar, Straus & Giroux, 2004.

Chace, James. *1912: Wilson, Roosevelt, Taft & Debs—The Election That Changed the Country.* New York: Simon & Schuster Paperbacks, 2004.

Chaffin, Tom. *Pathfinder: John Charles Fremont and the Course of American Empire*. New York: Hill & Wang, 2002.

Chernow, Ron. *Titan: The Life of John D. Rockefeller Sr.* New York: Random House, 1998.

Christensen, Clayton M. *The Innovator's Dilemma: When New Technologies Cause Great Firms to Fail*. Boston: Harvard Business School Press, 1997.

Christianson, Gale. *Greenhouse: The 200-Year History of Global Warming*. New York: Walker and Company, 1999.

Clymer, Adam. *Edward M. Kennedy: A Biography*. New York: William Morrow & Company, 1999.

Cohen, I. Bernard. *Benjamin Franklin: Scientist and Statesman*. New York: Charles Scribner's Sons, 1972.

———. *Science and the Founding Fathers*. Norton: New York, 1995.

Collier, Peter, and David Horowitz. *The Kennedys: An American Drama*. New York: Summit Books, 1984.

Colt, George Howe. *The Big House: A Century in the Life of an American Summer Home*. New York: Scribner, 2003.

Cowan, Edward. *Oil and Water: The Torrey Canyon Disaster*. Philadelphia and New York: J. B. Lippincott and Company, 1973.

Crile, George. "The Mellons, the Mafia and a Colonial County." *Washington Monthly*, June 1975.

Crocker, Zenas. *A History of Oyster Harbors to 1994*. Oyster Harbors, Osterville, MA: Oyster Harbors Property Owners Association Inc., 1994.

Cronkite, Walter. *A Reporter's Life*. New York: Alfred A. Knopf, 1996.

———. *Around America: A Tour of Our Magnificent Coastline*. New York: W. W. Norton and Company, 2001.

Crosby, Donald G. *Environmental Toxicology and Chemistry*. Oxford: Oxford University Press, 1998.

Damore, Leo. *The Cape Cod Years of John Fitzgerald Kennedy*. Englewood Cliffs, NJ: Prentice Hall, 1967.

Darley, Julian. *High Noon for Natural Gas: The New Energy Crisis*. White River Junction, VT: Chelsea Green Publishing, 2004.

Davis, Kenneth Sydney. *FDR, the New York Years, 1928–1933*. New York: Random House, 1979.

David, L. J. *Fleet Fire: Thomas Edison and the Pioneers of the Electric Revolution*. New York: Arcade Publishing, 2003.

Deffeyes, Kenneth S. *Beyond Oil: The View from Hubbert's Peak*. New York: Hill & Wang, 2005.

Deffeyes, Kenneth S. *Hubbert's Peak: The Impending World Oil Shortage*. Princeton, NJ: Princeton University Press, 2001.

De Waal, Frans. *Our Inner Ape*. New York: Riverside Books, 2005.

Drury, Allen. *Advise and Consent*. Garden City, NY: Doubleday, 1959.

Durden, Robert F. *Electrifying the Piedmont Carolinas: The Duke Power Company, 1904–1997*. Durham, NC: Carolina Academic Press, 2001.

Early, Tony. *Jim the Boy*. Boston: Little, Brown & Company, 2000.

"Edison's Electric Light: Conflicting Statements as to Its Utility." *New York Times*, October 21, 1879.

Emsley, John. *Nature's Building Blocks: An A-Z Guide to the Elements*. London: Oxford University Press, 2003.

Ettre, Leslie S. "M. S. Tswett and the Invention of Chromatography." *LCGC North America* 21, no. 5: 458.

Farrell, Jane. *The History of Cape Cod's Own . . . The Wianno Senior*. Self-published pamphlet. 1969.

Ferris, Timothy. *The Whole Shebang: A State-of-the-Universe(s) Report*. New York: Touchstone, 1998.

Feynman, Richard P. *Six Easy Pieces: Essentials of Physics Explained by Its Most Brilliant Teacher*. Cambridge, MA: Perseus Books, 1994.

———. *"What Do You Care What People Think?" Further Adventures of a Curious Character*. New York: Bantam Books, 1989.

Filler, Martin. "A Cool Mellon." *Vanity Fair*, April 1992.

Flavin, Christopher, and Nicholas Lenssen. *Power Surge: Guide to the Coming Energy Revolution*. New York: W. W. Norton, 1994.

Franklin, Benjamin. "Letter IV: From Benjamin Franklin Esq., in Philadelphia, to Peter Collinson, Esq., F.R.S. London: Farther Experiments and Observations in Electricity. 1748." In *Experiments and Observations on Electricity*. London: E. Cave, 1751. Digital edition accessed December 28, 2006, at Octavo Editions. http://www.octavo.com/editions/frkelc.

———. "Letter XI. From Benjamin Franklin, Esq; of Philadelphia. October 19, 1752." In *Experiments and Observations on Electricity*. London: E. Cave, 1751. Digital edition accessed December 28, 2006, at Octavo Editions. http://www.octavo.com/editions/frkelc.

Friedel, Robert. *Edison's Electric Light*. New Brunswick, NJ: Rutgers University Press, 1996.

Galison, Peter. *Einstein's Clocks, Poincare's Maps*. New York: W. W. Norton & Company, 2003.

Geisst, Charles R. *Wall Street: From Its Beginnings to the Fall of Enron*. New York: Oxford University Press, 2004.

Gelbspan, Ross. *Boiling Point: How Politicians, Big Oil and Coal, Journalists, and Activists Have Fueled the Climate Crisis—and What We Can Do to Avert Disaster*. New York: Basic Books, 2004.

———. *The Heat Is On: The Climate Crisis, the Cover-up, the Prescription*. Cambridge, MA: Perseus, 1998.

Ghosh, Amitav. *The Glass Palace*. New York: Random House, 2002.

Gibson, Barbara. *Life with Rose Kennedy: An Intimate Portrait*. New York: Warner, 1986.

Gibson, Robert. "A New Way of Light in the Valley." *Rural Electrification Magazine*, May 1986, 11–14.

———. "The Road to Chiniak: A meter a mile is dense enough in Alaska." *Rural Electrification Magazine*, April 1987, 18–23.

Giddens, Paul H. *Early Days of Oil: A Pictorial History of the Beginning of the Oil Industry*. Titusville, PA: The Colonel, Inc., 2000. (Reprinted from the Princeton University Press, 1948.)

Gleick, James. *Genius: The Life and Science of Richard Feynman*. New York: Pantheon, 1992.

———. *Isaac Newton*. New York: Vintage, 2004.

Goodell, Jeff. *Big Coal: The Dirty Secret Behind America's Energy Future*. Boston: Houghton Mifflin, 2006.

Goodwin, Doris Kearns. *The Fitzgeralds and the Kennedys: An American Saga*. New York: St. Martin's Press, 1987.

Grant, Robert. "Notes on the Pan-American Exposition," *Cosmopolitan Magazine*, September 1901.

Grossfeld, Stan, and David Halberstam. *Nantucket: The Other Season*. Chester, CT: Globe Pequot Press, 1982.

Hammond, John Winthrop. *Men and Volts: The Story of General Electric*. New York: J. B. Lippincott Company, 1941. (Copyright, General Electric.)

Hawthorne, Nathaniel. *House of the Seven Gables*. Boston and New York: Houghton Mifflin, 1952.

Heilbron, J. L. *Electricity in the 17th and 18th Centuries: A Study in Early Modern Physics*. Mineola, NY: Dover, 1999.

Hermann, C. David. *RFK: A Candid Biography of Robert F. Kennedy*. New York: Dutton, 1998.

Hightower, Jim. "Feeling Safer Yet?" *The Texas Observer*. October 12, 2001.

Hilty, James W. *Robert Kennedy: Brother Protector*. Philadelphia: Temple University Press, 1997.

Hirsch, Richard F. *Power Loss: The Origins of Deregulation and Restructuring in the American Electric Utility System*. Cambridge, MA: MIT Press, 1999.

Hoffman, Peter. *Tomorrow's Energy: Hydrogen, Fuel Cells and the Prospects for a Cleaner Planet*. Cambridge, MA: MIT Press, 2001.

Holmes, Hannah. *The Secret Life of Dust: From the Cosmos to the Kitchen Counter, the Big Consequences of Little Things*. New York: John Wiley & Sons Inc., 2001.

Horne, Alistair. *Seven Ages of Paris*. New York: Alfred A. Knopf, 2002.

Huber, Peter W., and Mark P. Mills. *The Bottomless Well: The Twilight of Fuel, the Virtue of Waste, and Why We Will Never Run Out of Energy*. New York: Basic Books, 2005.

Hughes, Thomas P. *Networks of Power: Electrification in Western Society, 1880–1930*. Baltimore: Johns Hopkins University Press, 1983.

———. *Rescuing Prometheus*. New York: Random House, 1998.

Hunt, John M. *Petroleum Geochemistry and Geology*. New York: W. H. Freeman & Company, 1995.

Israel, Paul. *Edison: A Life of Invention*. New York: John Wiley & Sons, 1998.

Jardine, Lisa. *The Curious Life of Robert Hooke: The Man Who Measured London*. New York: HarperCollins, 2004.

Josephson, Matthew. *Edison: A Biography*. New York: History Book Club, 2003.

Kapuscinski, Ryszard. *Shah of Shahs*. New York: Harcourt, Brace, & Jovanovich, 1992.

Keating, Paul W. *Lamps for a Brighter America*. New York: McGraw-Hill Book Company, Inc., 1954.

Kelley, Kitty. *Elizabeth Taylor: The Last Star*. New York: Simon & Schuster, 1981.

Kennedy, Edward M. *American Back On Track*. New York: Penguin, 2006.

Kennedy, Robert F., Jr. *Crimes Against Nature: How George W. Bush and His Corporate Pals Are Plundering the Country and Hijacking Our Democracy*. New York: HarperCollins, 2004.

Kirsch, David A. *The Electric Vehicle and the Burden of History*. New Brunswick, NJ and London: Rutgers University Press, 1998.

Klein, Edward. *Farewell, Jackie: A Portrait of Her Final Days*. New York: Viking, 2004.

Koskoff, David E. *The Mellons: The Chronicle of America's Richest Family*. New York: Thomas Y. Crowell Company, 1978.

Kristof, Nicholas D. "Island Nations Fear Sea Could Swamp Them." *New York Times*. December 1, 1997, sec. F, 9.

Kruglinski, Susan. "What's in a Gallon of Gas?" *Discover Magazine*, April 2004, 11.

Lane, Nick. *Oxygen: The Molecule That Made the World*. Oxford: Oxford University Press, 2002.

Larson, Erik. *The Devil in the White City: Murder, Magic and Madness at the Fair That Changed America*. New York: Crown, 2003.

Lawford, Christopher Kennedy. *Symptoms of Withdrawal*. New York: HarperCollins, 2005.

Leach, Robert J., and Peter Gow. *Quaker Nantucket: The Religious Community Behind the Whaling Empire*. Nantucket, MA: Mill Hill Press, 1997.

Leaming, Barbara. *Mrs. Kennedy: The Missing History of the Kennedy Years*. New York: Free Press, 2001.

Lee, Daniel Sheldon. *Buzzards Bay: A Journey of Discovery*. Beverly, MA: Commonwealth Editions, 2004.

Leggett, Jeremy. *The Carbon War: Global Warming and the End of the Oil Era*. New York: Routledge, 2001.

Leonard, Jonathan Norton. "People Come to Cape Cod." *The American Mercury*, November 1933, 295–302.

Littell, Robert T. *The Men We Became: My Friendship with John F. Kennedy Jr.* New York: St. Martin's Press, 2004.

Lovins, Amory B., E. Kyle Datta, Thomas Feiler, et al. *Small Is Profitable: The Hidden Economic Benefits of Making Electrical Resources the Right Size.* Snowmass, CO: The Rocky Mountain Institute, 2002.

Lungren, Charles M. "Electricity at the World's Fair." *The Popular Science Monthly.* October 1993.

Lynch, Michael P. *True to Life: Why Truth Matters.* Cambridge, MA: MIT Press, 2005.

McLean, Bethany, and Peter Elkind. *The Smartest Guys in the Room.* London and New York: Portfolio / Penguin, 2003.

McPhee, John. *Rising From the Plains.* New York: Farrar, Straus, Giroux, 1986.

———. *Uncommon Carriers.* New York: Farrar, Straus, & Giroux, 2006.

Mellon, Paul, with John Baskett. *Reflections in a Silver Spoon.* New York: William Morrow & Company, 1992.

Menand, Louis. *The Metaphysical Club: A Story of Ideas in America.* New York: Farrar Strauss & Giroux, 2002.

Meyer, Herbert W. *A History of Electricity and Magnetism.* Boston: MIT Press, 1971.

Meyer, Karl E. *The Dust of Empire: The Race for Mastery in the Asian Heartland.* New York: Public Affairs, 2003.

Mooney, Robert F. *Nantucket Only Yesterday: An Island View of the Twentieth Century.* Nantucket, MA: Wesco Press, 2000.

Moran, Richard. *Executioner's Current: Thomas Edison, George Westinghouse, and the Invention of the Electric Chair.* New York: Alfred A. Knopf, 2002.

Morris, Edmund. *Theodore Rex.* New York: Random House, 2001.

National Academy of Sciences. *Oil in the Sea.* Washington, DC: The National Academy Press, 1985.

———. *Oil in the Sea III: Inputs, Fates and Effects.* Washington, DC: The National Academy Press, 2003.

National Research Council. *Renewable Power Pathways: A Review of the U.S. Department of Energy's Renewable Energy Programs.* Washington, DC: National Academy Press, 2000.

Nicholl, Charles. *Leonardo da Vinci: Flights of the Mind.* New York: Viking, 2004.

Nye, David E. *Electrifying America: Social Meanings of a New Technology, 1880–1940.* Cambridge, MA: MIT Press, 1990.

———. *Consuming Power: A Social History of American Energies.* Cambridge, MA: MIT Press, 1996.

O'Connell, James C. *Becoming Cape Cod: Creating a Seaside Resort.* Lebanon, NH: University Press of New England, 2003.

Oppenheimer, Jerry. *The Other Mrs. Kennedy: Ethel Skakel Kennedy, An American Drama of Power, Privilege, and Politics*. New York: St. Martin's Press, 1994.

Pacala, Stephen, and R. Socolow. "Stabilization Wedges: Solving the Climate Problem for the Next Fifty Years with Current Technology," *Science* 305 (August 13, 2004): 968–72.

Pearson, Henry Greenleaf. *Son of New England: James Jackson Storrow*. Boston, MA: Privately printed by Helen Osborne Storrow, 1932.

Peck, Leonard W. *For Golden Friends I Had: An Account of a Lifelong Love Affair with Cotuit*. Osterville, MA: The Barnstable Book Company, 2000.

Pence, Richard A. Ed. *The Next Greatest Thing*. Washington, DC: The National Rural Electric Cooperative Association, 1984.

Perlin, John. *From Space to Earth: The Story of Solar Electricity*. Cambridge, MA: Harvard University Press, 2002.

———. "Solar Power: The Slow Revolution." *Inventions & Technology*, Summer 2002, 20–25.

Philbrick, Nathaniel. *Away Off Shore: Nantucket Island and Its People, 1602–1890*. Nantucket, MA: Mill Hill Press, 1994.

Picard, Lisa. *Dr. Johnson's London*. New York: St. Martin's Griffin, 2002.

Piel, Gerard. *The Age of Science: What Scientists Learned in the Twentieth Century*. New York: Basic Books, 2001.

Rainie, Harrison, and John Quinn. *Growing Up Kennedy: The Third Wave Comes of Age*. New York: G. P. Putnam's Sons, 1983.

Rand, Ayn. *The Fountainhead*. New York: Bobbs-Merrill Company Inc., 1968.

Revkin, Andrew. "Large Ice Shelf in Antarctica Disintegrates at Great Speed." *New York Times*, March 20, 2002, 1.

Roberts, Paul. *The End of Oil*. New York: Houghton Mifflin, 2004.

Rodgers, William. *Brown-Out: The Power Crisis in America*. New York: Stein & Day, 1972.

Robinson, Kim Stanley. *Red Mars, Green Mars, Blue Mars*. New York: Bantam, 1994.

Roosevelt, Theodore. *The Strenuous Life: Essays and Addresses*. New York: The Century Company, 1901.

Russell, Colin A. *Michael Faraday: Physics and Faith*. New York and Oxford: Oxford University Press, 2003.

Russell, Dick. *Striper Wars: An American Fish Story*. Washington, DC: Island Press, 2005.

Rybczynski, Witold. *A Clearing in the Distance: Frederick Law Olmsted and America in the 19th Century*. New York: Touchstone, 2000.

Schlesinger, Arthur M., Jr. *Robert Kennedy and His Times*. New York: Houghton Mifflin, 1978.

Schwartz, John. "Living on Internet Time, in Another Age." *New York Times*, April 4, 2002.

Seifer, Marc J. *The Life and Times of Nikola Tesla: Biography of a Genius*. New York: Citadel Press, 1998.

Sharlin, Harold Issadore. "Energy in Flowing Water and the Public Interest: Public and Private Power at Niagara Falls." For presentation at the American Historical Association in Dallas, December 1977. Document on file at the Hagley Museum and Library, Wilmington, DE.

Simon, Linda. *Dark Light: Electricity and Anxiety from the Telegraph to the X-Ray*. New York: Harcourt Inc., 2004.

Simon, Seymour. *Lightning*. New York: Morrow Junior Books, 1997.

Singh, Simon. *Big Bang: The Origin of the Universe*. New York: HarperCollins, 2004.

Slocum, Joshua. *Sailing Alone Around the World*. New York: Penguin Books, 1999.

Smil, Vaclav. *Enriching the Earth: Fritz Haber, Carl Bosch, and the Transformation of World Food Production*. Cambridge, MA: MIT Press, 2001.

Smith, Sally Bedell. *Grace and Power: The Private World of the Kennedy White House*. New York: Random House, 2004.

Taraborrelli, J. Randy. *Elizabeth*. New York: Warner Books, 2006.

Theroux, Paul. "Where Earth and Water Mix It Up," *Outside Magazine*, July 1999.

Thimmesch, Nick. "The Farmer Takes a Wife." *McCall's*, January 1977.

Thomas, Evan. *Robert Kennedy: His Life*. New York: Simon & Schuster, 2000.

Todd, Nancy Jack. *A Safe and Sustainable World: The Promise of Ecological Design*. Washington, DC: Island Press, 2005.

Tyndall, John. "The Electric Light." *The Popular Science Monthly*, March 1879.

Tyson, Neil de Grasse, Charles Liu, and Robert Irion. *One Universe: At Home in the Cosmos*. Washington, DC: Joseph Henry Press, 2000.

Tyson, Neil de Grasse, and Donald Goldsmith. *Origins: Fourteen Billion Years of Cosmic Evolution*. New York: W. W. Norton & Company, 2004.

Van Arsdale, John C. *It Wasn't My Time: The Autobiography of John C. Van Arsdale*. Privately published.

Victor, David G. *The Collapse of the Kyoto Protocol and the Struggle to Slow Global Warming*. Princeton, NJ: Princeton University Press, 2001.

Vaitheeswaran, Vijay V. *Power to the People: How the Coming Energy Revolution Will Transform an Industry, Change Our Lives, and May Even Save the Planet*. New York: Farrar, Straus & Giroux, 2003.

Verne, Jules. *Twenty Thousand Leagues Under the Sea*. New York: Charles Scribner's Sons, 1925.

Wald, Matthew L. "Company Hoping to Sell Fuel Cells: Hydrogen Maker May Aid in Producing Pollution-Free Cars." *New York Times*. March 14, 2002. Page C6, New England edition.

Ward, Diane Raines. *Water Wars: Drought, Flood, Folly and the Politics of Thirst*. New York: Riverhead Books, 2002.

Wertenbaker, William. "A Reporter at Large: A Small Spill." *The New Yorker*, November 26, 1973, 48–79.

Wianno Senior Class Association. *The Senior: 75 Years of the Wianno Senior Class*. Self-published, 1989.

Winchester, Simon. *The River at the Center of the World: A Journey Up the Yangtze, and Back in Chinese Time*. New York: Picador, 2004.

Wineapple, Brenda. *Hawthorne: A Life*. New York: Alfred A. Knopf, 2003.

Woodwell, George M. "Toxic Substances and Ecological Cycles." *Scientific American* 216, no. 3 (March 1967): 24–31.

Yardley, Jim. "China's Economic Engine Needs Power (Lots of It)." *New York Times*, March 14, 2004, 3.

Yergin, Daniel. *The Prize: The Epic Quest for Oil, Money & Power*. New York: Touchstone, 1993.

PublicAffairs is a publishing house founded in 1997. It is a tribute to the standards, values, and flair of three persons who have served as mentors to countless reporters, writers, editors, and book people of all kinds, including me.

I. F. STONE, proprietor of *I. F. Stone's Weekly*, combined a commitment to the First Amendment with entrepreneurial zeal and reporting skill and became one of the great independent journalists in American history. At the age of eighty, Izzy published *The Trial of Socrates*, which was a national bestseller. He wrote the book after he taught himself ancient Greek.

BENJAMIN C. BRADLEE was for nearly thirty years the charismatic editorial leader of *The Washington Post*. It was Ben who gave the *Post* the range and courage to pursue such historic issues as Watergate. He supported his reporters with a tenacity that made them fearless and it is no accident that so many became authors of influential, best-selling books.

ROBERT L. BERNSTEIN, the chief executive of Random House for more than a quarter century, guided one of the nation's premier publishing houses. Bob was personally responsible for many books of political dissent and argument that challenged tyranny around the globe. He is also the founder and longtime chair of Human Rights Watch, one of the most respected human rights organizations in the world.

• • •

For fifty years, the banner of Public Affairs Press was carried by its owner Morris B. Schnapper, who published Gandhi, Nasser, Toynbee, Truman, and about 1,500 other authors. In 1983, Schnapper was described by *The Washington Post* as "a redoubtable gadfly." His legacy will endure in the books to come.

Peter Osnos, *Founder and Editor-at-Large*